S0-ADZ-114

Proceedings in Life Sciences

Milton H. Stetson
Editor

Processing of Environmental Information in Vertebrates

With Contributions by
S. BINKLEY, C.L. BROWN, P. DEVICHE, R.S. DONHAM,
J.A. ELLIOTT, B.D. GOLDMAN, R.L. GOODMAN, C.A. MARLER,
F.L. MOORE, M.C. MOORE, S.M. REPPERT, M.H. STETSON,
H. UNDERWOOD, M. WATSON-WHITMYRE, D.R. WEAVER,
F.E. WILSON, J.C. WINGFIELD

With 81 Figures

Springer-Verlag
New York Berlin Heidelberg
London Paris Tokyo

Milton H. Stetson
School of Life and
  Health Sciences
University of Delaware
Newark, Delaware 19706
USA

QP
84
.6
.P77
1988

Library of Congress Cataloging-in-Publication Data
Processing of environmental information in
    vertebrates.
    (Proceedings in life sciences)
    Includes bibliographies and index.
    1. Biological rhythms. 2. Photoperiodism.
3. Vertebrates—Physiology. I. Stetson, Milton H.
II. Series.
QP84.6.P77 1987   596'.0188   87-12671

© 1988 by Springer-Verlag New York Inc.
All rights reserved. This work may not be translated or copied in whole or in part without the
written permission of the publisher (Springer-Verlag, 175 Fifth Avenue, New York, NY 10010,
USA), except for brief excerpts in connection with reviews or scholarly analysis. Use in connection
with any form of information storage and retrieval, electronic adaptation, computer software, or
by similar or dissimilar methodology now known or hereafter developed is forbidden.
The use of general descriptive names, trade names, trademarks, etc. in this publication, even if
the former are not especially identified, is not to be taken as a sign that such names, as understood
by the Trade Marks and Merchandise Marks Act, may accordingly be used freely by anyone.

Typeset by David E. Seham Associates, Inc., Metuchen, New Jersey.
Printed and bound by Arcata Graphics/Halliday, West Hanover, Massachusetts.
Printed in the United States of America.

9 8 7 6 5 4 3 2 1

ISBN 0-387-96558-0 Springer-Verlag New York Berlin Heidelberg
ISBN 3-540-96558-0 Springer-Verlag Berlin Heidelberg New York

*Dedicated to Professor Donald S. Farner
in honor of his 70th birthday*

Preface

This book arises from a symposium by the same name presented to the Division of Comparative Endocrinology of the American Society of Zoologists. The major objective was to demonstrate neural, neuroendocrine, and endocrine mechanisms by which vertebrates translate and transduce environmental cues into precise physiological responses. To the extent that we have succeeded, credit is due the contributors. Their cooperation and that of the editorial staff of Springer-Verlag, New York, has made my job much easier. Financial support from the National Science Foundation (DCB 85-07866) is gratefully acknowledged.

Newark, Delaware   Milton H. Stetson

Contents

Contributors

S. Binkley, Biology Department, Temple University, Philadelphia, Pennsylvania, U.S.A.

C.L. Brown, Department of Zoology, University of California, Berkeley, California, U.S.A.

P. Deviche, Department of Zoology, Oregon State University, Corvallis, Oregon, U.S.A.

R.S. Donham, School of Life and Health Sciences, University of Delaware, Newark, Delaware, U.S.A.

J.A. Elliott, Worcester Foundation for Experimental Biology, Shrewsbury, Massachusetts, U.S.A.

B.D. Goldman, Worcester Foundation for Experimental Biology, Shrewsbury, Massachusetts, U.S.A.

R.L. Goodman, Department of Physiology, West Virginia University Medical Center, Morgantown, West Virginia, U.S.A.

C.A. Marler, Department of Zoology, Arizona State University, Tempe, Arizona, U.S.A.

F.L. Moore, Department of Zoology, Oregon State University, Corvallis, Oregon, U.S.A.

M.C. Moore, Department of Zoology, Arizona State University, Tempe, Arizona, U.S.A.

S.M. Reppert, Children's Service, Massachusetts General Hospital, Boston, Massachusetts, U.S.A.

M.H. Stetson, School of Life and Health Sciences, University of Delaware, Newark, Delaware, U.S.A.

H. Underwood, Department of Zoology, North Carolina State University, Raleigh, North Carolina, U.S.A.

M. Watson-Whitmyre, School of Life and Health Sciences, University of Delaware, Newark, Delaware, U.S.A.

D.R. Weaver, Children's Service, Massachusetts General Hospital, Boston, Massachusetts, U.S.A.

F.E. Wilson, Division of Biology, Ackert Hall, Kansas State University, Manhattan, Kansas, U.S.A.

J.C. Wingfield, Department of Zoology, University of Washington, Seattle, Washington, U.S.A.

Chapter 1

Photoperiod Effects on Thyroid Function in Fish

I. Introduction

The hypothesis that thyroid function in fish may be altered by photoperiod arose, in part, from observations of seasonal changes in thyroid activity. Cyclic changes in thyroid function follow a variety of patterns in teleosts, and they remain generally poorly understood in terms of how they are regulated and what physiological purposes they serve. In some instances, seasonal alterations in thyroid function coincide with sexual development or temperature changes, but in others it appears that photoperiod could be the primary environmental cue for thyroid seasonality. Correlative changes alone only indirectly infer cause-and-effect relationships, however, and controlled experiments have been required in order to establish the specific environmental variables that govern thyroid function. Experimentation with fish maintained on controlled lighting has shown that a change in daylength alone can have profound effects on thyroid function in teleosts, although it has become abundantly clear that temperature and other factors are also important. Species differences appear to be the rule rather than the exception.

There are also reports of diurnal rhythms in thyroid activity in a few teleost species, although it is not yet known how prevalent these cycles are or what regulates them. The superimposition of daily rhythms on seasonal rhythms could conceivably account for some of the complicated patterns of thyroid activity that have been observed. Compounded rhythms of this sort may modulate the production and peripheral actions of prolactin in teleosts (Sage and deVlaming 1975). At this time, however, it would be unwise to assume that diurnal thyroid cycles are ubiquitous among teleosts.

In this report, seasonal and diurnal cycles in thyroid function in teleost fish are examined as they may relate to photoperiodic control mechanisms. The interpretation of the biological significance of these cycles is limited by the plentiful gaps and contradictions in the literature on physiological roles of thyroid hormones in fish. The experimental evidence implicating photoperiodic regulation of thyroid activity in fish is summarized, and one means by which this environmental variable could exert influences at the level of the thyroid follicles is also considered.

II. Seasonal Changes in Thyroid Function

Eales (1979), in his comprehensive review of thyroid function in fish, pointed out that although seasonal changes in thyroid activity have been documented in more than 20 teleost species, few if any sweeping generalizations could safely be made. One recurring theme in this body of literature is a spring or summer peak in thyroid activity followed by late-summer quiescence, but Eales (1979) noted that these seasonal patterns appeared to be complicated by the simultaneous effects of temperature, photoperiod, and certain kinds of physiological changes.

In the earliest of investigations of thyroid seasonality in fish, qualitative differences in the histological appearances of thyroid follicles were described. Citing changes in the size and shape of thyroid follicular epithelial cells and in the appearance of colloid in Atlantic salmon, *Salmo salar,* Hoar (1939) concluded that increased glandular activity was evident in spring, with signs of moderate activity in summer and minimal activity in winter. This pattern was comparable to those previously found in the loach, *Misgurnus fossilis,* by Lieber (1936) and in the eel, *Anguilla vulgaris,* by von Hagen (1936). The agreement among these studies and earlier work on amphibians and reptiles led Hoar (1939) to summarize that thyroid function in ectotherms is not directly related to temperature as it is (inversely) in homeotherms.

Barrington and Matty (1955) described the consistent effects of thyrotropin treatment on thyroid follicular epithelial cell height and applied this quantitative technique to the study of thyroid seasonality in *Phoxinus phoxinus.* The peak in epithelial cell height in this minnow occurred in February, although a rather sparse sampling protocol was used. It was determined experimentally that thyroid epithelial cell height was positively correlated with temperature under conditions that included constant darkness, suggesting to these authors that the midwinter activation of the thyroid took place despite a presumably inhibitory effect of low temperatures (Barrington and Matty 1955). Similar histological cyclicity was discovered in the epithelial cell height measurements of both immature and adult cod (Woodhead 1959). These results were in direct contrast to the earlier reports, and they set a precedent by (1) establishing the need to consider species differences in thyroid seasonality and (2) raising the possibility of environmental variables other than temperature in the regulation of these cycles.

A new dimension in the analysis of thyroid activity was added with the advent of radiochemical methods, which have been used widely in studies of thyroid function in fish since the mid-1950s. Iodine-131 uptake was found to vary seasonally in *Fundulus heteroclitus,* with maximal uptake in January, although this result was difficult to reconcile with a sharp peak in radiothyroxin production that spanned the summer months (Berg et al. 1959). This was perhaps the first case in which multiple indicators of thyroid activity in fish seemed to produce contradictory results, which led Berg et al. (1959) to consider radioiodine uptake, thyroid hormone synthesis, and secretion as independent events. Apparent discrepancies between various radiochemical and histological criteria have since become more commonplace (Eales 1965, 1979).

Swift (1960) examined environmental factors that could be involved in the regulation of a seasonal cycle in radioiodine loss from the thyroid glands of yearling brown trout *(Salmo trutta)*. This cycle was followed over a two-year period, and was characterized by highest rates of radioiodine loss, presumably indicating rapid hormone turnover, in winter and early spring. A second increase in radioiodine loss, of relatively lower magnitude, occurred during the spawning seasons (July) in both years. Swift (1960) postulated that two sets of regulatory mechanisms were driving this cycle: (1) a seasonal response to temperature and/or photoperiod and (2) an ancillary mechanism reflecting involvement of thyroid hormones in gonadal maturation during the reproductive season. Inverse correlation of temperature with iodine loss appeared somewhat stronger than the correlation of daylength with iodine loss, and Swift (1960) concluded that thyroid function probably increased in winter in response to cooler temperatures. Neither photoperiod nor temperature correlated with iodine loss in a simple, linear fashion, however, and there is little experimental evidence for temperature compensation in teleost thyroid function.

Swift's (1960) idea of a primary seasonal thyroid cycle and a secondary activation coincident with some aspect or aspects of spawning was reinforced in a later study in which plasma thyroxin levels were measured in samples collected at approximately weekly intervals throughout the spawning season in brown trout (Pickering and Christie 1981). A distinct elevation of plasma thyroxin (T_4) was detected in female trout over the four weeks during which ovulation took place, but significant changes were not seen in the males. In general, biphasic thyroid cycles appear to be present in many teleosts, although the relative degree of activation during the reproductive season is highly variable between species (reviewed by Leatherland 1982). In the char *(Salvelinus leucomaenis pluvius)*, histological indicators changed in a cyclic pattern with both spring and late-summer (spawning seson) peaks (Honma and Tamura 1965). Thyroid follicular cell height in the goby *(Leukogobius guttatus)* paralleled the reproductive cycle; in this case, peak thickness was coincident with spawning and was followed by a precipitous decline and then a recovery in the winter months (Tamura and Honma 1973). In a feral population of bullheads *(Ictalurus nebulosus)*, circulating levels of T_4 and triiodothyronine (T_3) appeared to be positively correlated with temperature and daylength, with the exception of a sudden spike in T_4 in April, a time of increasing photoperiod, temperature, and gonadal recrudescence (Burke and Leatherland 1983). Yaron (1969) described pronounced seasonal changes in thyroid epithelial cell height in the cyprinid fish, *Acanthobrama terrae-sanctae*. This was a biphasic cycle that appeared to be loosely and positively correlated with temperature (and presumably daylength) throughout most of the year, but with a distinct peak during the winter spawning period.

Anadromous salmonids exhibit changes in thyroid status that are associated with smoltification (Hoar 1939; reviewed by Folmar and Dickhoff 1980), but little is known of the specific actions of thyroid hormones in this process. A distinct surge in plasma T_4 is generally seen in fish undergoing or about to undergo smoltification (Dickhoff et al. 1978). The rise in plasma T_4 in coho salmon (Oncorhynchus kisutch) may be caused, in part, by an increase in thyroidal sensitivity to thyrotropin (Specker and Richman 1984) and is accompanied by increases in both T_4 secretion and metabolic clearance rates (Specker et al. 1984). There is some evidence that photoperiod-thyroid interactions could be of central importance in the coordination of smoltification-related events. Experimental manipulations of photoperiod can alter the timing of smoltification (Saunders and Henderson 1970; reviewed by Wedemeyer et al. 1980), probably, in part, by advancing or delaying the characteristic T_4 surge (Grau et al. 1982). The control of smoltification does not appear to be a simple matter, however. Environmental cues other than photoperiod (notably the lunar cycle and temperature) appear to be involved (Grau et al. 1981, 1982) and hormones other than thyroid hormones are needed to initiate the process. Treatment of yearling Atlantic salmon with T_3 promoted some changes that occur during smoltification (e.g., increased seawater tolerance) but had no effect on others (body lipid content and gill sodium-potassium ATPase activity; Saunders et al. 1985). Likewise, treatment of underyearling amago salmon (Oncorhynchus rhodurus) with T_4 alone produced fewer smoltification-related changes than treatment with T_4 and growth hormone together (Miwa and Inui 1985).

It is clear that no single pattern can be said to exemplify the seasonal variations described in the teleost thyroid literature (Table 1-1). Aside from peaks that occur coincidentally with gonadal maturation, or in the case of salmonids, with or prior to smoltification, the reported cycles in teleost thyroid activity have peaks at several different times of year. Most often, these seem to occur in spring. The variations in the reported thyroid cycles undoubtedly reflect both species differences and differences in techniques. Many of the apparent differences could be a consequence of the reliance of most authors on one or two indicators of thyroid activity, despite the inherent limitations of this approach (discussed by Eales 1979). Disagreement among indicators underscores the fallacy in the notion that all aspects of thyroid function change in unison. Secretion of thyroid-stimulating hormone (TSH), radioiodine uptake, and thyroid epithelial cell height can all increase in the course of a systemic surge in thyroid hormone production and utilization, but these changes are also seen in some hypothyroid states. Likewise, circulating thyroid hormone levels alone are an inadequate measure of thyroid system function: the concentrations of T_3 and T_4 in plasma are determined by both secretion and peripheral turnover rates and are therefore subject to possible misinterpretation. Thus, the term "thyroid activity" may be overly simplistic as it is often used, and the reevaluation of some of these seasonal patterns may be necessary.

Table 1-1. Peaks in Thyroid Activity of Teleosts

Genus	Means of assessment	Time(s) of peak	Citation
Salmo	Qualitative histol.	Spring-summer	Hoar 1939
Salmo	Radioiodine loss Thyroid epith. ht.	Feb.-Mar., July	Swift 1955, 1959, 1960
Salmo	Serum T_3 and T_4	Nov., Apr.	Osborn et al. 1978
Salmo	Conv. ratio	Variable	Eales 1963, 1965
Oncorhynchus	Thyroid epith. ht.	Mostly in May	Eales 1963, 1965
Salvelinus	Thyroid epith. ht.	Spring, summer	Honma and Tamura 1965
Salvelinus	Serum T_3 and T_4	Apr.-May	White and Henderson 1977
Salvelinus	Serum T_4	Apr.	McCormick and Naiman 1984
Phoxinus	Thyroid epith. ht.	Feb.-May	Barrington and Matty 1955
Misgurnus	Qualitative histol.	Spring-summer	Lieber 1936
Anguilla	Qualitative histol.	Spring-summer	von Hagen 1936
Gaddus	Thyroid epith. ht.	Dec.-Jan.	Woodhead 1959
Fundulus	^{131}I uptake ^{131}I T_4 prod.	Jan. June	Berg et al. 1959
Perca	TSH bioassay	Feb. (?)	Swift and Pickford 1965
Mystus	^{131}I uptake	Apr.-May	Singh 1968b,c
Mystus	TSH bioassay	Apr.-May	Singh and Sathyanesan 1968
Acanthobrama	Thyroid epith. ht.	Feb., Aug.	Yaron 1969
Alosa	TSH cell nucl. dia.	Mar.	Li 1969
Leukogobius	Thyroid epith. ht.	June	Tamura and Honma 1973
Heteropneustes	Thyroid epith. ht.	June	Pandey and Munshi 1976
Pleuronectes	Serum T_3 and T_4	Summer, winter	Osborn and Simpson 1978
Pseudopleuronectes	Serum T_3 Serum T_4	Fall-winter Apr.-June	Eales and Fletcher 1982
Ictalurus	Serum T_3 Serum T_4	Aug. Apr.	Burke and Leatherland 1983
Fundulus	Serum T_4, post-TSH treatment	Jan.	Grau et al. 1985

III. Diurnal Rhythms in Thyroid Function

Most investigators of seasonal endocrine cycles (and other endocrine phenomena) take the precaution of collecting all of their samples at the same time of day, in order to minimize the influences of possible diurnal variations. If, in fact, there are daily rhythms in thyroid hormone secretion or utilization, it is important to characterize any such patterns and determine the environmental and/or physiological factors which are capable of entraining them. There is little consensus on the subject of diurnal rhythmicity of teleost thyroid function despite a substantial amount of work in salmonids and in goldfish. White and Henderson (1977) found that T_4 levels in brook trout *(Salvelinus fontinalis)* serum were lower in samples collected at 0530 than at either 1130 or 2030, which hinted at a daily rhythm. Further support for such a rhythm was obtained when significant variations in plasma T_4 in brook trout were reported by McCormick and Naiman (1984). Leatherland et al. (1977) found no differences in rainbow trout *(Salmo gairdneri)* plasma T_4 related to the time of sampling, although their samples covered only half of the 24-hour period. More thorough sampling schedules were followed in the investigations of Brown et al. (1978) and Osborn et al. (1978), although these authors reached opposite conclusions about the presence or absence of diurnal patterns in serum T_4 in the rainbow trout. In one study, serum T_4 levels did not vary significantly in samples collected at ten times over a 48-hour period (Brown et al. 1978). In contrast, Osborn et al. (1978) reported diurnal patterns in serum T_4 at two different times of year. Osborn et al. (1978) determined that T_4 levels were suppressed during daylight hours and elevated during darkness, although a clear pattern was not obvious in the raw T_4 data. Log-transformed plasma T_4 data subjected to a curve-smoothing statistical treatment (four-point moving averaging) were interpreted as evidence for a marked diurnal rhythm in the rainbow trout. No such rhythms have been reported in subsequent studies of circulating thyroid hormones in unfed rainbow trout (Milne and Leatherland 1979, 1980), but a diel rhythm in plasma T_3 and T_4 was found in rainbow trout that were fed shortly before experimentation (Eales et al. 1981). The importance of feeding in diurnal thyroid rhythmicity in rainbow trout was further investigated by Flood and Eales (1983), who found transient plasma T_3 and T_4 elevations four hours after feeding previously starved trout. It has been proposed that diurnal T_4 rhythms may change seasonally or may differ among populations within a given species (Noeske and Spieler 1983). Dietary variations may be responsible for some seasonal changes in thyroid rhythms. For example, there was no clear diurnal rhythm in plasma T_4 in Atlantic salmon in March (when the fish were not feeding), but a distinct rhythm was seen in May, after feeding began and water temperatures increased (Rydevik et al. 1984).

Diurnal rhythms in plasma T_3 and T_4 have been reported in goldfish *(Carassius auratus)* held on twelve hours of light and twelve hours of darkness (12L:12D), with significant peaks of both hormones occurring eight hours after the onset of light (Spieler and Noeske 1979). Essentially the same thyroxin pattern was reproduced in goldfish on two different feeding schedules, but the existence of a rhythm in circulating T_3 was dependent on the time of feeding (Spieler and Noeske 1981). The possibility that a thyroid rhythm in goldfish may be entrained by daylength has been investigated by measuring T_4 in goldfish acclimated for one month to controlled photoperiods (Noeske and Spieler 1983). The experimental photoperiods included 8L:16D, 16L:8D, and two 12L:12D lighting regimes, with lights-on occurring at different times of day. The T_4 rhythm was abolished under 8L:16D and under one set of 12L:12D lighting conditions, and peaks were coincident with the onset of light in the other two groups. It is of interest to note that peak T_4 levels in this experiment (about 150 ng/100 ml; Noeske and Spieler 1983) were more than an order of magnitude lower than the approximate levels reported earlier (about 4000 ng/100 ml; Spieler and Noeske 1979). The apparent differences in the magnitude and phase relationships of serum T_4 peaks seen in goldfish under 12L:12D lighting in these experiments suggest that either light alone does not entrain this diurnal rhythm or one month is not sufficient time for entrainment.

IV. Regulation of Thyroid Function by Photoperiod and Other Variables

The hormonal output of the teleostean thyroid gland is subject to several kinds of regulation. It is a typical vertebrate thyroid system in the sense that thyroid hormone synthesis and secretion are governed primarily by TSH and can be influenced by negative feedback (see reviews by Gorbman 1969, Eales 1979). The hypothalamic regulation of pituitary TSH production in teleosts is unique among vertebrates for at least two reasons: hypothalamic regulation of TSH is mainly inhibitory in nature and it appears to be exerted by neural rather than humoral means (see Stetson and Grau 1980). Also, a wide variety of endocrine compounds have been proposed to have modulatory effects on the pituitary-thyroid axis and on the peripheral actions of thyroid hormones in fish. Gonadotropic and gonadal hormones, growth hormone, prolactin, and other compounds are frequently mentioned in this context (reviewed by Leatherland 1982) and could be involved in thyroid seasonality. Injection of ovine prolactin, for example, causes a reduction of serum T_4 levels in *Fundulus heteroclitus* (Grau and Stetson 1977) by impairment of secretion rather than an effect on peripheral T_4 clearance (Brown and Stetson 1983). Ovine prolactin increases peripheral deiodination of T_4 to T_3 in the eel, *Anguilla anguilla* (Leloup and de Luze 1980), but salmon prolactin has no such effect (de Luze and Leloup 1984). The relative importance of prolactin and other potential hormonal modulators of thyroid function cannot be ascertained without a critical reexamination of the reported effects of heterologous hormones and continued study of the actions of homologous and endogenous hormones.

In more than half of the seasonal patterns reported, there is some suggestion of thyroid activation during the reproductive season. As Leatherland (1982) has pointed out, the cause of a change in thyroid status is particularly difficult to pinpoint during a spring spawning season, when so many environmental and physiological changes are occurring simultaneously. It is possible that spawning-related thyroid secretion could be stimulated by gonadotropins (see below) or gonadal steroids (Singh 1968a, van Overbeeke and McBride 1971) as well as by seasonal changes, including photoperiod. A study by McCormick and Naiman (1984) addressed the issue of photoperiodic control of thyroid and reproductive cycles in the brook trout. In this experiment, plasma T_4 was measured in fish held on natural and three-month-delayed photoperiods. Spawning was delayed by three months in the latter group, but the effect of the shifted photoperiod on the T_4 cycle was not clear. Nevertheless, spawning in both groups was associated with elevated plasma T_4 levels.

The possible involvement of thyroid hormones in teleost reproduction has often been suggested, although the evidence linking the two endocrine systems is, for the most part, indirect (Leatherland 1982). The major exception is a series of studies of the freshwater perch *(Anabas testudineus),* in which changes in circulating thyroid hormones are tightly synchronized with gonadal development (Chakraborti and Battacharya 1984). Experimental treatments with ovarian steroids and some gonadotropins caused significant increases in thyroid hormone levels (Chakraborti et al. 1983, Chakraborti and Battacharya 1984), which raises the possibility that reproductive hormones may be involved in the regulation of thyroid activity and its coordination with spawning in this species. These results must be interpreted carefully, however. The actions of mammalian or partially purified salmonid gonadotropins on the teleost thyroid do not necessarily infer a physiological role, since such actions may differ from those of endogenous gonadotropins (Brown et al. 1985). Thyroid hormone treatment sensitizes the ovary to some actions of gonadotropins in the freshwater perch (Sen and Battacharya 1981) and in the stellate sturgeon, *Acipenser stellatus* (Detlaff and Davydova 1979). The recent demonstrations of high-affinity T_3 binding sites in ovarian nuclei and the promotion of protein synthesis by T_3 in ovarian tissue strengthen the argument for thyroid involvement in the reproduction of the freshwater perch (Chakraborti et al. 1986).

Experiments involving the manipulation of photoperiod have demonstrated that fish usually respond with altered thyroid function, among other responses. Fish have been acclimated to constant darkness, with generally inhibitory effects on thyroid function. These conditions probably produce the most physiologically meaningful results for the blind cave fishes and their relatives. Exposure of *Astyanax mexicanus* (the sighted variety) to constant darkness caused hypertrophy of the thyroid tissue and, later, thyroid atrophy (Rasquin 1949, Rasquin and Rosenbloom 1954). A biphasic seasonal pattern in pituitary TSH content in the catfish *Mystus vittatus* was disrupted under conditions of constant darkness and the cycle was advanced in constant light (Singh 1967, Singh and Sathyanesan 1968). These lighting conditions had similar effects on the cyclic pattern of iodine-131 uptake in the same species of catfish (Singh 1968b,c).

Several groups of investigators have acclimated fish to artificial photoperiods in order to compare the effects of relatively long or short days. One trend has emerged in a majority of these studies, albeit not so convincingly in some reports as in others: acclimation to short days tends to promote thyroid activity and acclimation to long days tends to have inhibitory effects. These studies began with the work of Hoar and Robertson (1959), who reported that thyroid radioiodine uptake was slightly (although not significantly) increased in goldfish held on an 8L:16D cycle, compared to fish held on a 16L:8D cycle. These results were repeated in a later study in which both radioiodine uptake and T_4-T_3 conversion ratios seemed to increase in response to acclimation to an LD 8:16 photoperiod, but again the trend suggesting a difference to the authors was not substantiated statistically (Hoar and Eales 1963). It was also noted that goldfish held on the shorter photoperiod had better chill resistance than those on the long photoperiod and that TSH treatment could improve chill resistance, but the evidence for short-day stimulation of the thyroid and a role of thyroid hormones in cold tolerance was circumstantial (Hoar and Eales 1963).

The effects of four different photoperiods (8L:16D, 16L:8D, daylength increasing from 8L:16D to 16L:8D, and daylength decreasing from 16L:8D to 8L:16D) on one indicator of thyroid activity have been studied in the green sunfish (Lepomis cyanellus). Radioiodine loss from the thyroid region increased under the short photoperiods and decreased under the longer ones (Gross et al. 1963).

Honma (1966) examined thyroid histology in a captive stock of ayu (Plecoglossus altivelis) after having exposed the fish for one year to artificially lengthened days by switching on lights for two to three hours per night. He found that thyroid degeneration was prevalent in the surviving stock, although the high mortality rate and the lack of a control group in this experiment must be considered in the interpretation of these results.

Short-day stimulation of thyroid function in fish is by no means an inflexible rule, however. Indirect evidence of long-day stimulation of the thyroid was obtained by Baggerman (1960), who correlated thyroid activity with salinity preference in yearling coho salmon and three-spined sticklebacks (Gasterosteus aculeatus). She found that increased daylength caused a change in the behavioral orientation to salinity, and she cited radiochemical and other indications that thyroid activity increased just before the onset of migratory behavior. Based on these results, she postulated that photoperiod might "release" migratory behavior partly by way of thyroid activation, and subsequent studies tend to support her argument (Fontaine 1975, Youngson and Simpson 1984). It is also worthwhile to note that T_4 surges associated with smoltification can occur in either spring or fall, depending on the species (Nishioka et al. 1985). These observations raise questions regarding the apparently contradictory stimulatory effects of both long and short photoperiods on teleost thyroid function. Long or short photoperiods, per se, may be less important than the relative rate of change in daylength (Wagner 1974), and there are indications that the responses of fish to increasing or decreasing photoperiods can vary at different times of year (Clarke et al. 1978). The photoperiodic history of fish prior to experimentation (which is seldom reported) may be another important variable; the initial change in daylength encountered in the adaptation to an artificial photoperiod could conceivably be more important than the absolute daylength.

M.H. Stetson and coworkers have investigated the control of seasonal changes in thyroid function in *Fundulus heteroclitus*. In this euryhaline teleost, basal serum T_4 levels did not change significantly with the seasons but did appear to be more variable in winter (Grau et al. 1985). The serum T_4 elevations following ovine TSH (oTSH) treatment, however, were much greater in fish collected in winter than in those collected in summer, despite the facts that (1) these experiments were conducted at the same temperature and (2) thyroid hormone clearance rates were similar in summer and winter. The oTSH responsiveness of this teleost changed in a cyclic pattern over a two-year period, with the maximal response in midwinter and the minimal response in summer, and this cycle was inversely correlated with daylength ($p < 0.001$) but not temperature (Figure 1-1, after Grau et al. 1985).

The environmental regulation of this cycle was then investigated. A test of the thyroidal response to oTSH after one month of exposure to artificial short days (8L:16D) or long days (14L:10D) demonstrated significantly greater serum T_4 elevations in the short-day-acclimated fish (Figure 1-2, after Grau et al. 1985). Experimentally, reduced water temperature both lowered basal T_4 and impaired the responsiveness of the thyroid gland to oTSH. This suggested that the enhanced responsiveness to oTSH seen in winter was opposite what might be expected taking the short-term thermal effects alone into account. Grau et al. (1985) proposed that a photoperiod-induced increase in thyroid responsiveness served to counteract winter temperatures, in order to aid in the year-round maintenance of thyroid function.

Further investigations of seasonal differences in the serum T_4 response of *Fundulus* to oTSH revealed sharply contrasting time courses (Figure 1-3; Brown and Stetson 1985). Treatment with oTSH caused sustained elevations in winter, which took several days to return to basal levels. In summer, isolated T_4 peaks were evident, and there was some oscillation after the last oTSH injection. We interpreted these data as an indication of possible seasonal differences in sensitivity to negative feedback, the persistence of hormone elevation in

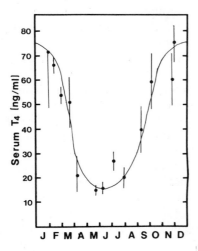

Figure 1-1. Serum T_4 elevations in *Fundulus heteroclitus* following four daily oTSH injections. There was a highly significant inverse correlation of daylength with serum T_4 after oTSH treatment. Redrawn with permission from Grau et al. 1985.

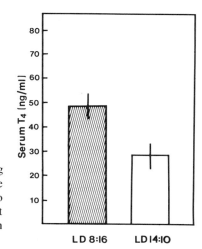

Figure 1-2. One month of exposure to a long (14L:10D) photoperiod significantly reduced the thyroidal response of *Fundulus heteroclitus* to oTSH, as compared to the effects of a short (8L:16D) photoperiod. Redrawn with permission from Grau et al. 1985.

winter suggesting minimal feedback sensitivity as opposed to the summer pattern, which suggested more acute sensitivity. The sensitivity of the thyroids of summer- and winter-caught *Fundulus* to negative feedback was tested by measuring the effects of immersion in graded doses of thyroid hormones on radioiodine uptake, and it was confirmed that sensitivity was greater in summer. Inhibitory effects occurred at lower concentrations of T_3 and T_4 in fish captured in summer (Figure 1-4, after Brown and Stetson 1985). Given the strong inverse correlation of the response to oTSH with environmental photoperiod (Figure 1-1) and the duplication of these changes with artificially controlled photoperiods (Figure 1-2), we then tested the hypothesis that the observed seasonal differences in the negative feedback properties of thyroid hormones might be under pho-

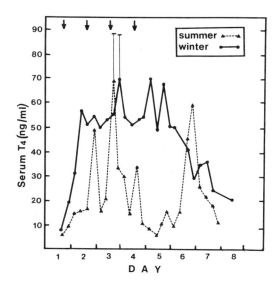

Figure 1-3. Seasonal differences in the time courses of response to oTSH (four daily injections, indicated by arrows). The largest standard errors of the mean (vertical bars) occurred on the third day of both summer and winter experiments. Redrawn with permission from Brown and Stetson 1985.

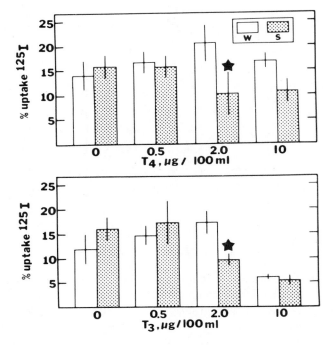

Figure 1-4. The inhibitory effects of T_4 (top panel) and T_3 (bottom panel) on radioiodine uptake in winter (open bars) and summer (shaded bars). Both hormones had significant feedback effects at a dosage of $2\mu g/100$ ml aquarium water in summer, but these effects were absent in winter. Redrawn with permission from Brown and Stetson 1985.

toperiodic control. Fish were held for one month on artificial photoperiods (8L:16D and 14L:10D, as before) and the effects of intermediate doses of thyroid hormones on radioiodine uptake were measured. The results (Figure 1-5) show an enhancement of feedback sensitivity to T_4 in fish acclimated to the long (14L:10D) photoperiod.

In summary, these results illustrate a mechanism by which a change in the environment can induce an adjustment of thyroid function of possible adaptive value. A short-day-induced reduction in negative feedback sensitivity could promote prolonged responses to TSH, with the net effect of maintaining plasma thyroid hormone levels at a time when temperature effects would tend to inhibit secretion. The opposite effect, increased feedback in response to lengthening days, could contribute to the attenuation of responses to TSH in the summer, when thermal effects might tend to promote hypersecretion. We have suggested that photoperiodic control of negative feedback properties may be an important aspect of the environmental regulation of thyroid function in *Fundulus heteroclitus*, and we have speculated that this could be useful in counteracting adverse temperature effects. There are numerous questions that are yet to be addressed. It is not known whether this mechanism is present in other species; this pos-

Figure 1-5. A negative feedback effect of T_4 (10 μg/100 ml aquarium water) on radioiodine uptake was seen in fish held on long days (14L:10D) but was absent in fish held on short days (8L:16D). Redrawn with permission from Brown and Stetson 1985.

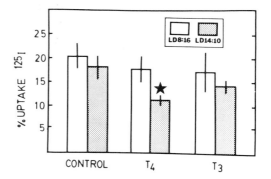

sibility merits further consideration and could help explain the generally stimulatory effects of short days on teleost thyroid function. Also, it has not yet been determined how a change in daylength could affect negative feedback. The regulation of hypothalamic and pituitary thyroid hormone receptors would be the obvious place to start. Finally, the assessment of the ultimate value of a regulatory mechanism of this sort must await a more thorough understanding of the physiological roles of thyroid hormones in teleosts as related to seasonally changing conditions.

Acknowledgments. Thanks are expressed to the organizers of the symposium (Dr. Milton Stetson, in particular) and to Michael Greenblatt and Dr. Stephen McCormick for their thoughtful and constructive reviews of the manuscript.

References

Baggerman B (1960) Factors in the diadromous migrations of fish. Symp Zool Soc Lond 1:33–60.

Barrington EJW, Matty AJ (1955) Seasonal variation in the thyroid gland of the minnow, *Phoxinus phoxinus* L., with some observations on the effect of temperature. Proc Zool Soc Lond 124:89–95.

Berg O, Gorbman A, Kobayashi H (1959) The thyroid hormones in invertebrates and lower vertebrates. In: Gorbman A (ed) Comparative Endocrinology, pp. 302–319. John Wiley and Sons, New York.

Brown CL, Stetson MH (1983) Prolactin-thyroid interaction in *Fundulus heteroclitus.* Gen Comp Endocrinol 50:167–171.

Brown CL, Stetson MH (1985) Photoperiod-dependent negative feedback effects of thyroid hormones in *Fundulus heteroclitus.* Gen Comp Endocrinol 58:186–191.

Brown CL, Grau EG, Stetson MH (1985) Functional specificity of gonadotropin and thyrotropin in *Fundulus heteroclitus.* Gen Comp Endocrinol 58:252–258.

Brown J, Fedoruk K, Eales JG (1978) Physical injury due to injections or blood removal causes transitory elevations of plasma thyroxine in rainbow trout, *Salmo gairdneri.* Can J Zool 56:1998–2003.

Burke M, Leatherland JF (1983) Seasonal changes in serum thyroid hormone levels of the feral brown bullhead, *Ictalurus nebulosus* Lesueur. J Fish Biol 23:585–593.

Chakraborti P, Battacharya S (1984) Plasma thyroxine levels in freshwater perch: influence of season, gonadotropins, and gonadal hormones. Gen Comp Endocrinol 53:179–186.

Chakraborti P, Rakshit DK, Battacharya S (1983) Influence of season, gonadotropins, and gonadal hormones on the thyroid activity of freshwater perch, *Anabas testudineus* (Bloch). Can J Zool 61:359–364.

Chakraborti P, Maitra G, Battacharya S (1986) Binding of thyroid hormones to isolated ovarian nuclei from a freshwater perch, *Anabas testudineus*. Gen Comp Endocrinol 62:239–246.

Clarke WC, Shelbourn JE, Brett JR (1978) Growth and adaptation to sea water in "underyearling" sockeye *(Oncorhynchus nerka)* and coho *(O. kisutch)* salmon subjected to regimes of constant or changing temperature and day length. Can J Zool 56:2413–2421.

Detlaff TA, Davydova SI (1979) Differential sensitivity of cells of follicular epithelium and oocytes in the stellate sturgeon to unfavorable conditions, and correlating influence of triiodothyronine. Gen Comp Endocrinol 39:236–243.

Dickhoff WW, Folmar LC, Gorbman A (1978) Changes in plasma thyroxine during smoltification of coho salmon, *Oncorhynchus kisutch*. Gen Comp Endocrinol 36:229–232.

Eales JG (1963) A comparative study of thyroid function in migrant juvenile salmon. Can J Zool 41:811–824.

Eales JG (1965) Factors influencing seasonal changes in thyroid activity in juvenile steelhead trout, *Salmo gairdneri*. Can J Zool 43:719–729.

Eales JG (1979) Thyroid functions in cyclostomes and fishes. In: Barrington EJW (ed) Hormones and Evolution, pp. 341–436. Academic Press, New York.

Eales JG, Fletcher GL (1982) Circannual cycles of thyroid hormones in plasma of winter flounder (*Pseudopleuronectes americanus* Walbaum). Can J Zool 60:304–309.

Eales JG, Hughes M, Uin L (1981) Effect of food intake on diel variation in plasma thyroid hormone levels in rainbow trout, *Salmo gairdneri*. Gen Comp Endocrinol 45:167–174.

Flood CG, Eales JG (1983) Effects of starvation and refeeding on plasma T_4 and T_3 levels and T_4 deiodination in rainbow trout, *Salmo gairdneri*. Can J Zool 61:1949–1953.

Folmar LC, Dickhoff WW (1980) Parr-smolt transformation (smoltification) and seawater adaptation in salmonids; a review of selected literature. Aquaculture 21:1–37.

Fontaine M (1975) Physiological mechanisms in the migration of marine and amphihaline fish. Adv Mar Biol 13:241–355.

Gorbman A (1969) Thyroid function and its control in fishes. In: Hoar WS, Randall DJ (eds) Fish Physiology, Vol II, pp. 241–274. Academic Press, New York.

Grau EG, Stetson, MH (1977) The effects of prolactin and TSH on thyroid function in *Fundulus heteroclitus*. Gen Comp Endocrinol 33:329–335.

Grau EG, Dickhoff WW, Nishioka RS, Bern HA, Folmar LC (1981) Lunar phasing of the thyroxine surge preparatory to seaward migration of salmonid fish. Science 211:607–609.

Grau EG, Specker JL, Nishioka RS, Bern HA (1982) Factors determining the occurrence of the surge in thyroid activity in salmon during smoltification. Aquaculture 28:49–58.

Grau EG, Brown CL, Stetson MH (1985) Photoperiodic regulation of thyroid responsiveness to TSH in *Fundulus heteroclitus*. J Exp Zool 234:199–205.

Gross WL, Fromm PO, Roelofs EW (1963) Relationship between thyroid and growth in green sunfish, *Lepomis cyanellus*. Trans Am Fish Soc 92:401–407.

Hoar WS (1939) The thyroid gland of the Atlantic salmon. J Morphol 65:257–295.

Hoar WS, Eales JG (1963) The thyroid gland and low temperature resistance of goldfish. Can J Zool 41:653–669.

Hoar WS, Robertson GB (1959) Temperature resistance of goldfish maintained under controlled photoperiods. Can J Zool 37:419–428.

Honma Y (1966) Studies on the endocrine glands of a salmonid fish, the ayu, *Plecoglossus altivelis* Temminck et Schlegel. VI. Effect of artificial controlled light on the endocrines of the pond-cultured fish. Bull Jpn Soc Sci Fish 32:32–40.

Honma Y, Tamura E (1965) Studies on the Japanese chars, the iwana (genus *Salvelinus*). I. Seasonal changes in the endocrine glands of the Nikko-iwana, *Salvelinus leucomaenis pluvius* (Hilgendorf). Bull Jpn Soc Sci Fish 31:867–877.

Leatherland JF (1982) Environmental physiology of the teleostean thyroid gland: a review. Env Biol Fish 7:83–110.

Leatherland JF, Cho CY, Slinger SJ (1977) Effects of diet, ambient temperature and holding conditions on plasma thyroxine levels in rainbow trout *Salmo gairdneri*. J Fish Res Bd Can 34:677–682.

Leloup J, de Luze A (1980) Prolactine et hormones thyroïdiennes chez l'anguille (*Anguilla anguilla* L.). C R Acad Sci (Paris) 291:87–90.

Li AH (1969) Cyclic changes of the gonadotrophs, thyrotrophs and corticotrophs in the adenohypophysis of the alewife, *Alosa pseudoharengus* (Wilson). MS thesis, University of Wisconsin-Milwaukee.

Lieber A (1936) Cited by Hoar (1939).

de Luze A, Leloup J (1984) Fish growth hormone enhances peripheral conversion of thyroxine to triiodothyronine in the eel (*Anguilla anguilla* L.). Gen Comp Endocrinol 56:308–312.

McCormick SD, Naiman RJ (1984) Osmoregulation in the brook trout, *Salvelinus fontinalis*—I. diel, photoperiod and growth related physiological changes in freshwater. Comp Biochem Physiol 79A:7–16.

Milne RS, Leatherland JF (1979) Temporal effects of pituitary hormone preparations on plasma thyroid hormone concentrations in rainbow trout. J Endocrinol 85:37P–38P.

Milne RS, Leatherland JF (1980) Changes in plasma thyroid hormones following administration of exogenous pituitary hormones and steroid hormones to rainbow trout (*Salmo gairdneri*). Comp Biochem Physiol 66A:679–686.

Miwa S, Inui Y (1985) Effects of L-thyroxine and ovine growth hormone on smoltification of amago salmon (*Oncorhynchus rhodurus*). Gen Comp Endocrinol 58:436–442.

Nishioka RS, Young G, Bern HA, Jochimsen W, Hiser C (1985) Attempts to intensify the thyroxin surge in coho and king salmon by chemical stimulation. Aquaculture 45:215–225.

Noeske TA, Spieler RE, (1983) Photoperiod and diel variations of serum cortisol, thyroxine and protein in goldfish, *Carassius auratus* L. J Fish Biol 23:705–710.

Osborn RH, Simpson TH (1978) Seasonal changes in thyroidal status in the plaice, *Pleuronectes platessa* L. J Fish Biol 12:519–526.

Osborn RH, Simpson TH, Youngson AF (1978) Seasonal and diurnal rhythms of thyroidal status in the rainbow trout, *Salmo gairdneri* Richardson. J Fish Biol 12:531–540.

Pandey BN, Munshi JSD (1976) Role of the thyroid in regulation of metabolic rate in an air-breathing siluroid fish, *Heteropneustes fossilis* (Bloch) J Endocrinol 69:421–425.

Pickering AD, Christie P (1981) Changes in the concentrations of plasma cortisol and thyroxine during sexual maturation of the hatchery-reared brown trout, *Salmo trutta* L. Gen Comp Endocrinol 44:487–496.

Rasquin P (1949) The influence of light and darkness on thyroid and pituitary activity of the characin *Astyanax mexicanus* and its cave derivatives. Bull Am Mus Nat Hist 93:497–528.

Rasquin P, Rosenbloom L (1954) Endocrine imbalance and tissue hyperplasia in teleosts maintained in darkness. Bull Am Mus Nat Hist 104:359–426.

Rydevik M, Lindahl K, Fridberg G (1984) Diel pattern of plasma T_3 and T_4 levels in Baltic salmon parr (*Salmo salar* L.) during two seasons. Can J Zool 62:643–646.

Sage M, deVlaming VL (1975) Seasonal changes in prolactin physiology. Am Zool 15:917–922.

Saunders RL, Henderson EB (1970) Influence of photoperiod on smolt development and growth of Atlantic salmon (*Salmo salar*). J Fish Res Bd Can 27:1295–1311.

Saunders RL, McCormick SD, Henderson EB, Eales JG, Johnston CE (1985) The effect of orally administered 3,5,3'-triiodo-L-thyronine on growth and salinity tolerance of Atlantic salmon (*Salmo salar* L.). Aquaculture 45:143–156.

Sen S, Battacharya S (1981) Role of thyroxine and gonadotropins on the mobilization of ovarian cholesterol in a teleost *Anabas testudineus* (Bloch). Indian J Exp Biol 19:408–412.

Singh TP (1967) Influence of photoperiods on the seasonal fluctuations of TSH content of the pituitary in a freshwater catfish, *Mystus vittatus* (Bloch). Experientia 23:1016–1017.

Singh TP (1968a) Thyroidal ^{131}I uptake and TSH potency of the pituitary in response to graded doses of methyltestosterone in *Mystus vittatus* (Bloch). Gen Comp Endocrinol 11:1–4.

Singh TP (1968b) Effects of varied photoperiods on rhythmic activity of thyroid gland in a teleost *Mystus vittatus* (Bloch). Experientia 24:93–94.

Singh TP (1968c) Seasonal changes in radioiodine uptake and epithelial cell height of the thyroid gland in freshwater teleosts *Esomus dandricus* (Ham.) and *Mystus vittatus* (Bloch) under varying conditions of illumination. Z Zellforsch 87:422–428.

Singh TP, Sathyanesan AG (1968) Thyroid activity in relation to pituitary thyrotropin level and their seasonal changes under normal and varied photoperiods in the freshwater teleost *Mystus vittatus* (Bloch). Acta Zool, Stockholm 54:47–56.

Specker JL, Richman NH (1984) Environmental salinity and the thyroidal response to thyrotropin in juvenile coho salmon (*Oncorhynchus kisutch*). J Exp Zool 230:329–333.

Specker JL, DiStephano JJ, Grau EG, Nishioka RS, Bern HA (1984) Development-associated changes in thyroxine kinetics in juvenile salmon. Endocrinology 115:399–406.

Spieler RE, Noeske TA (1979) Diel variations in circulating levels of triiodothyronine and thyroxine in goldfish, *Carassius auratus*. Can J Zool 57:665–669.

Spieler RE, Noeske TA (1981) Timing of a single daily meal and diel variations of serum thyroxine, triiodothyronine, and cortisol in goldfish, *Carassius auratus*. Life Sciences 28:2939–2944.

Stetson MH, Grau EG (1980) Hypothalamo-adenohypophysial relationships among vertebrates. In: Pang PKT, Epple A (eds) Evolution of vertebrate endocrine systems. Texas Tech Univ Press, Lubbock, Texas.

Swift DR (1955) Seasonal variations in the growth rate, thyroid gland activity and food reserves of brown trout (*Salmo trutta* Linn.). J Exp Biol 32:751–764.

Swift DR (1959) Seasonal variation in the activity of the thyroid gland activity of yearling brown trout (*Salmo trutta* Linn.). J Exp Biol 32:751–764.

Swift DR (1960) Cyclical activity of the thyroid gland of fish in relation to environmental changes. Symp Zool Soc Lond 2:17–27.

Swift DR, Pickford GE (1965) Seasonal variations in the hormone content of the pituitary gland of the perch, *Perca fluviatilis* L. Gen Comp Endocrinol 5:354–365.

Tamura E, Honma Y (1973) Histological changes in the organs and tissues of Gobiid fishes throughout its life span—V. Seasonal changes in the branchial organs of the flat-head goby in relation to sexual maturity. Bull Jpn Soc Sci Fish 39:1003–1011.

van Overbeeke AP, McBride JR (1971) Histological effects of 11-keto-testosterone, 17-α-methyltestosterone, estradiol, estradiol cypionate, and cortisol on the interrenal tissue, thyroid gland, and pituitary gland of gonadectomized sockeye salmon *(Oncorhynchus nerka)*. J Fish Res Bd Can 28:477–484.

von Hagen F (1936) Cited by Hoar (1939).

Wagner HH (1974) Photoperiod and temperature regulation of smolting in steelhead trout *Salmo gairdneri*. Can J Zool 52:219–234.

Wedemeyer GA, Saunders RL, Clarke WC (1980) Environmental factors affecting smoltification and early marine survival of anadromous salmonids. Mar Fish Rev 42:1–14.

White BA, Henderson NE (1977) Annual variations in circulating levels of thyroid hormones in brook trout, *Salvelinus fontinalis*, as measured by radioimmunoassay. Can J Zool 55:475–481.

Woodhead AD (1959) Variations in the activity of the thyroid gland of the cod, *Gaddus callarius* L., in relation to its migration in the Barents Sea. I. Seasonal changes. J Mar Biol Assoc UK 38:407–415.

Yaron Z (1969) Correlation between spawning, water temperature, and thyroid activity in *Acanthobrama terrae-sanctae* (Cyprinidae) of lake Tiberias. Gen Comp Endocrinol 12:604–608.

Youngson AF, Simpson TH (1984) Changes in serum thyroxine levels during smolting in captive and wild Atlantic salmon, *Salmo salar* L. J Fish Biol 24:29–39.

Chapter 2

Neuroendocrine Processing of Environmental Information in Amphibians

Frank L. Moore and Pierre Deviche

I. Amphibian Reproduction

A. Introduction

To survive and reproduce, animals must process environmental information, then respond appropriately. Some responses are rapid, such as predator avoidance; others are long-term, such as seasonal cycles in reproduction. Many such responses have been studied in amphibians, more than can be reviewed here. Primitive amphibians, as ancestors to reptiles, birds, and mammals, were the first vertebrates to live on land. Given their evolutionary position, an understanding of how amphibians process environmental information can provide insights into basic neuroendocrine mechanisms of other vertebrates. The present chapter focuses on the neuroendocrine controllers of reproduction that are associated with long-term annual reproductive cycles and with short-term changes in reproductive behaviors. The chapter does not review the neurophysiological studies pertaining to how amphibians locate and discriminate auditory and visual stimuli (for reviews of these topics, see Ewert et al. 1983, Ingle and Crews 1985).

Amphibians exhibit a wide variety of life history patterns and inhabit a broad range of environments (Duellman and Trueb 1986). Despite this diversity, amphibians are limited by the availability of water. Even those plethodontid salamanders which are entirely terrestrial require a surface film of water and high humidity for successful reproduction. Others, such as *Necturus* and *Amphiuma*, are permanently aquatic and have requirements like those of many fishes. Most amphibians, as their name implies, occupy terrestrial habitats as adults and only return to water to mate and lay eggs. Species that are adapted to seasonally stable environments, such as cave-dwelling salamanders (*Typhlomolge rathbuni*) and a few neotropical species (*Rana erythraea* and *Bolitoglossa subpalmata*), can mate during any month. Most amphibians, however, live in seasonally variable environments and have precise cycles of reproduction.

Most anurans (frogs and toads) reproduce by external fertilization. Mating in many anurans is initiated when the male captures a female in a dorsal amplectic clasp, gripping the female with the forelegs. After a variable period of time in amplexus—depending on species, ambient factors, and reproductive conditions—eggs and sperm are synchronously released into the water. The only anurans that have internal fertilization belong to the genera *Ascaphus* and *Nectophrynoides*.

Most urodeles (newts and salamanders) reproduce by internal fertilization. Mating usually involves a series of courtship behaviors and the subsequent transfer of sperm into the female's cloaca by means of a spermatophore (a package of sperm). For example, after completing the amplectic phase of courtship, rough-skinned newts (*Taricha granulosa*) use a reciprocal sequence of stereotyped behaviors that allows the male to deposit a spermatophore on the bottom of the pond and the female to pick it up in her cloaca (Moore et al. 1979a). Once in the cloaca, sperm move into specialized ducts, the so-called spermatotheca, that function to prolong sperm viability. In *T. granulosa*, ovulation usually occurs several weeks after insemination, and the eggs are fertilized when they pass through the lower oviducts or cloaca. This capacity to store sperm, which in some species may be for several years, permits the acts of mating and egg laying to be temporally discrete events.

B. Environmental Synchronizers

Three environmental factors primarily influence reproduction in amphibians: photoperiod, water (rainfall and humidity), and temperature (reviewed by Salthe and Mecham 1974, Jørgensen et al. 1978, Lofts 1984).

The effects of light duration (photoperiod) on amphibian reproduction appear to be less pronounced than the photoperiodic responses of many birds and mammals. However, in at least some species of amphibians, experiments demonstrate a clear effect of photoperiod on gametogenesis. For example, in *Plethodon cinereus*, spermatogenesis is accelerated by photoperiods of 16L:8D in early spring (Werner 1969). Likewise, in *Rana esculenta*, active spermatogenesis is maintained with a photoperiod of at least 12L:12D (Rastogi et al. 1978). In contrast, other studies have not found any effect of photoperiod on amphibian gametogenesis (Lofts 1984). Overall, it appears that in amphibians, photoperiod does not function as a dominant activator of gonadal regression or recrudescence, although it appears to modulate or synchronize gametogenesis in some species.

Since amphibians are poikilothermic, it is not surprising that amphibian gonadal activity is influenced by temperature. In temperate climates, most amphibians mate and lay eggs during late winter to early summer, a time of increasing ambient temperatures. Low ambient temperatures during winter or summer can disrupt spermatogenesis and ovarian growth (Jørgensen et al. 1978, Lofts 1984). Even in *R. esculenta* and *P. cinereus*, photoperiodically influenced species, low ambient temperatures can retard or arrest spermatogenesis. Conversely, spermatogenesis in these two species is enhanced by photoperiod only when ambient temperatures are elevated (cited above). In *T. granulosa*, males change from aquatic, breeding condition to nonbreeding, terrestrial condition more rapidly if held at warmer temperatures (Brown et al. 1984).

In summary, the above studies provide insights into the role of temperature in amphibian reproduction. First, gametogenesis and reproductive development can be accelerated or retarded by changes in ambient temperature. Second, seasonal cycles in reproduction can occur even when amphibians are held at constant temperatures. Thus, changes in ambient temperatures apparently can synchronize reproductive events but are not necessary for driving seasonal breeding cycles. This helps to explain why year-to-year variations in the onset of warm, vernal temperatures are correlated with the onset of breeding for particular amphibians at particular sites (Salthe and Mecham 1974).

The third environmental factor that affects amphibian reproduction is water. Amphibians need enough rainfall for the establishment of breeding ponds, moist substrate, and/or high humidity. Unlike light and temperature, water does not appear to influence the annual cycles in gonadal development. Instead, rainfall and high humidity are associated with the activation of migration to the breeding sites and the initiation of courtship and mating (Salthe and Mecham 1974). Extreme examples come from species that occupy xeric habitats and are opportunistic breeders, such as spadefoot toads (*Scaphiopus* spp.) that emerge and breed within a day or two after heavy rains.

C. Seasonal Changes in Reproduction

Annual changes in reproduction have been studied in many amphibians, providing a wealth of information about changes in the gonads, secondary sexual characteristics, and sexual behaviors (Lofts 1984). Marked changes in gonad weights occur in many species (see examples of seasonal changes in testicular weights in Figure 2-1). The increases in gonad weights reflect several different phenomena: the multiplication and growth of germ cells, the increase in testicular water content at the time of spermiation, and the uptake of yolk by the mature oocytes. The decreases, on the other hand, result from the release of mature sperm and eggs during the breeding season and from the resorption of germinal and nongerminal tissues after completion of breeding. Because gonadal development is influenced by environmental factors, the exact timing of gametogenesis varies with the geographic region, climatic condition, and species of amphibian.

Associated with seasonal cycles in gonadal development are changes in secondary sexual characteristics. In anurans, the most pronounced of these are the nuptial pads, which are pigmented callosities on the fingers or toes that assist the male in holding a female in amplexus. Examples of secondary sexual characteristics in male urodeles include nuptial pads, enlarged caudal tail fins, sexually dimorphic coloration, and enlarged cloacal glands. In seasonally breeding species, these structures are more tightly correlated to breeding activity

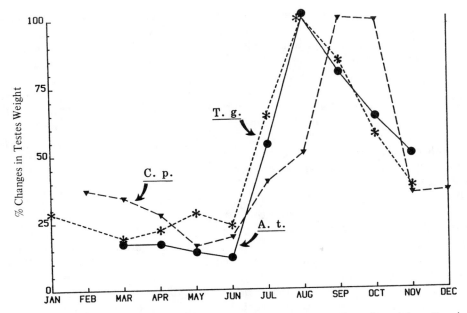

Figure 2-1. Seasonal changes in testicular weights in three species of urodeles. *C.p.* is *Cynops pyrrhogaster*, from Tanaka and Iwasawa 1979; *T.g.* is *Taricha granulosa*, from Specker and Moore 1980; *A.t.* is *Ambystoma tigrinum*, from Norris et al. 1985. Percent testes weights are calculated from maximum value for each species (*C.p.*, 105 mg; *T.g.*, 315 mg; *A.t.*, 1.64 g).

than are testicular weights (Figure 2-2), mainly because small testes during the breeding season are caused by the evacuation of sperm from the testes into the vas deferens (Specker and Moore 1980, Norris et al. 1985).

Reproductive behaviors also are restricted to particular environmental conditions or seasons. As a general rule, mating and egg-laying behaviors of temperate-zone amphibians occur from late winter to early summer. In those amphibians with external fertilization, males and females exhibit reproductive behaviors when environmental conditions simultaneously favor successful mating and embryonic development. In contrast, in those urodeles with internal fertilization and long-term sperm storage, mating and ovulation can occur at different seasons and under different conditions. In *Bolitoglossa rostrata*, a typical neotropical salamander (Houck 1977), mating occurs year-round. The males have continuous spermatogenesis and persistent secondary sexual characteristics whereas females have a discrete seasonal cycle of ovarian development, with ovulation and oviposition occurring in November or December. In rough-skinned newts (*T. granulosa*), most mating behaviors are mostly observed in February, March, and April whereas egg-laying behaviors are observed in May and June (Specker and Moore 1980). Therefore, in these urodeles, mating behaviors and egg-laying behaviors may be under different selective pressures and may be activated by different environmental conditions.

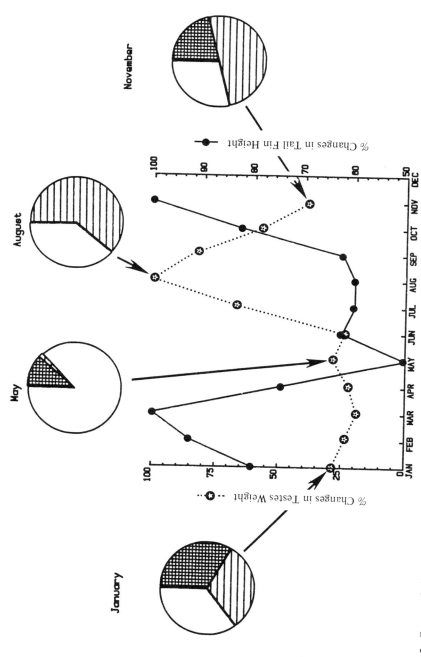

Figure 2-2. Seasonal changes in tail fin height, testicular weight, and testis composition in *Taricha granulosa* (data from Specker and Moore 1980). Percent tail fin height and testes weight are based on maximum values of 20 mm and 315 mg, respectively. Testis com- position represents the proportion of testis containing spermatogonia and spermatocytes (horizontal bars), spermatids and spermatozoa (cross-hatching), and evacuated testicular lobules (blank).

II. Gonadal Steroids and Gonadotropins

A. Testicular Steroid Hormones

Seasonal cycles in amphibian reproduction are correlated with seasonal changes in testicular steroid hormones. The predominant testicular steroids in the plasma of male amphibians are testosterone and 5α-dihydrotestosterone, but lower concentrations of other steroids also have been detected, namely, 11-ketotestosterone, 11β-hydroxytestosterone, androstenedione, progesterone, 17β-estradiol, and estrone (Bolaffi et al. 1979, Licht et al. 1983, Garnier 1985). Although in vitro studies confirm that testosterone is secreted by the testes (Moore et al. 1979b), some of the testosterone in the plasma appears to originate from nontesticular tissues (Moore et al. 1983).

In some amphibians, dihydrotestosterone concentrations exceed those of testosterone. For example in *Rana catesbeiana*, about equal titers of each are found in males during the winter, and up to four times greater concentrations of dihydrotestosterone than testosterone are found in males during the spring (Licht et al. 1983). In other species, testosterone concentrations are two or more times higher than dihydrotestosterone concentrations (Moore and Muller 1977, Norris et al. 1985, Siboulet 1981). Although the physiological importance of the different androgen ratios has not been proven, they probably relate to the types of receptors and enzyme systems (e.g., aromatase and 5α-reductase) in peripheral tissues (Callard et al. 1978).

All males for which there are measurements of plasma androgen concentrations are seasonal breeders and have pronounced seasonal changes in plasma androgen concentrations (Figure 2-3). In *T. granulosa*, plasma androgen concentrations begin to increase during late summer, reaching peak levels in late fall or winter (October or February, depending on year), and then decrease precipitously in early spring (Specker and Moore 1980, Zoeller and Moore 1985). The pattern of seasonal change for plasma androgens is similar for the other urodeles that have been studied (D'Istria et al. 1974, Clergue et al. 1985, Garnier 1985, Cayrol et al. 1985, Imai et al. 1985, Lecouteux et al. 1985, Norris et al. 1985). Other species, particularly *R. catesbeiana* (Licht et al. 1983), have different patterns of plasma androgen concentrations (Figure 2-3). The differences in plasma androgen profiles is perhaps related to the fact that bullfrogs are adapted to warmer climates and, in the translocated populations that were studied, have late breeding seasons.

Testosterone and other gonadal steroids are transported in the blood of amphibians in close association with carrier proteins, the sexual steroid binding proteins (SBP). These proteins have high affinity and limited capacity for binding gonadal steroids (Moore et al. 1983, Santa-Coloma et al. 1985). The movement of steroid hormones into target tissues, for example, the movement of steroids across the blood–brain barrier into the brain, is thought to be influenced by whether or not the steroid is bound to SBP. In *T. granulosa*, when androgen titers are high, most of the androgen is not bound to SBP; in contrast, when androgen titers are low, most androgen is bound to SBP (Moore et al. 1983),

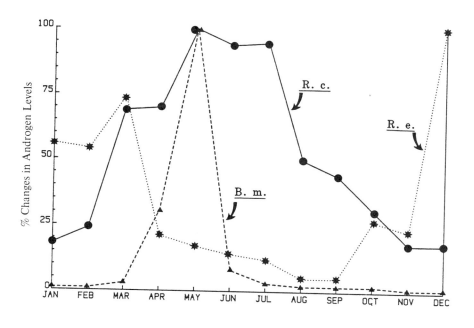

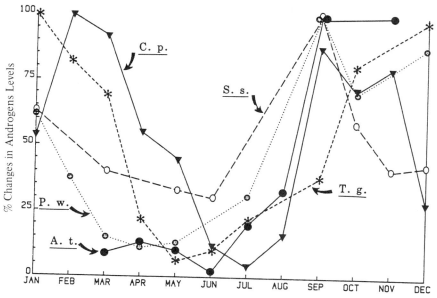

Figure 2-3. Seasonal variations in plasma androgen concentrations in anurans (upper panel) and urodeles (lower panel). *R.c.* is *Rana catesbeiana*, from Licht et al. 1983; *R.e.* is *Rana esculenta*, from D'Istria et al. 1974; *B.m.* is *Bufo mauritanicus*, from Siboulet 1981; *C.p.* is *Cynops pyrrhogaster*, from Imai et al. 1985; *S.s.* is *Salamandra salamandra*, from Lecouteux et al. 1985; *P.w.* is *Pleurodeles waltl*, from Cayrol et al. 1985; *A.t.* is *Ambystoma tigrinum*, from Norris et al. 1985; *T.g.* is *Taricha granulosa*, from Zoeller and Moore 1985. Data represent plasma concentrations of testosterone (*B.m.* and *R.e.*), dihydrotestosterone (*R.c.*), or total androgen immunoreactivity (other species). Percent androgen level was calculated from maximum values (ng/ml) for each species (*R.c.*, 27; *R.e.*, 19; *B.m.*, 595; *C.p.*, 65; *S.s.*, 107; *P.w.*, 58; *A.t.*, 39; *T.g.*, 38).

indicating that there is a seasonal change in the ratio of bound to free plasma steroids.

B. Ovarian Steroid Hormones

Female amphibians exhibit seasonal changes in ovarian steroids. The predominant ovarian steroids in the plasma of female amphibians are testosterone, androstenedione, 17β-estradiol, estrone, and progesterone (Lofts 1984, Pierantoni et al. 1984). In female *R. catesbeiana* and *R. esculenta*, plasma testosterone and estradiol concentrations generally are associated with ovarian development, such that they reach peak levels at the time of ovulation and then decrease precipitously after ovulation (Licht et al. 1983, Polzonetti-Magni et al. 1984). The physiological importance of testosterone in female amphibians is not known.

C. Functions of Gonadal Steroids

Seasonal changes in gonadal steroid hormones cause, in part, the pronounced seasonal changes in secondary sex characteristics. For example, plasma androgen concentrations in male amphibians are correlated with the development of morphological structures such as nuptial pads, caudal tail fins, and cloacal glands (Specker and Moore 1980, Garnier 1985, Norris et al. 1985). Gonadal steroids also are important to amphibian reproductive behaviors (reviewed by Kelley and Pfaff 1978, Moore 1983). They are involved in the control of male and female sexual behaviors in several species. In female *Xenopus laevis*, sexual receptivity can be restored in ovariectomized animals with implants of estrogen plus progesterone but not with implants of either estrogen or progesterone alone (Kelley 1982). In *Triturus cristatus*, *X. laevis*, and *T. granulosa* males, sexual behaviors are abolished by castration and are restored by replacement treatments with testosterone and/or dihydrotestosterone (Kelley and Pfaff 1976, Moore 1978, Andreoletti et al. 1983). These studies provide good evidence that gonadal steroids are necessary for the expression of sexual behaviors in these amphibians; however, there also is evidence that gonadal steroids are not the only hormones involved in the regulation of amphibian sexual behaviors (reviewed in Moore 1983, 1986). For example, on any given date, the occurrence of sexual behaviors is not correlated with plasma androgen concentrations (Moore and Muller 1977). Temperate-zone amphibians exhibit reproductive behaviors during the time of, or shortly after, high plasma androgen concentrations (Specker and Moore 1980, Mendonca et al. 1985, Norris et al. 1985). Therefore, high plasma androgen titers and sexual behaviors are not always temporally associated (see Crews 1984). For this reason and others (see Moore, 1983), it appears that testosterone and other testicular steroids control amphibian sexual behaviors by preparing the animal to express the behavior, not by controlling day-to-day changes in the propensity to express the behavior.

Gonadal steroids appear to regulate sexual behaviors by acting on a variety of target cells, not by acting exclusively on some steroid-sensitive "switch" in the brain (Kelley 1980). Gonadal steroids influence neurons in the central control centers in the brain plus in the motor and sensory pathways. There is evidence that androgens can affect cells in the region of the anterior preoptic nucleus (Wada and Gorbman 1977). There also is electrophysiological evidence that testosterone can directly affect neurons (Schmidt 1980). In *X. laevis*, dihydrotestosterone binding sites exist in auditory pathways (torus semicircularis), motor pathways (dorsal tegmental area), and central hypothalamic areas (e.g., anterior preoptic area), all of which are involved in sexual behaviors (Kelley 1980). Dihydrotestosterone binding sites occur in the spinal cord in areas that are involved in the control of clasping behaviors (Erulkar et al. 1981). Furthermore, androgen treatments can increase electrophysiological activity in motoneurons that are involved in clasping behaviors (Erulkar et al. 1981).

D. Gonadotropins

Two chemically distinct glycoprotein hormones, follicle-stimulating hormone (FSH) and luteinizing hormone (LH), have been extracted and purified from *R. catesbeiana* and other amphibians (Licht 1979). In contrast, another laboratory has extracted and purified a single gonadotropin with both LH and FSH biological activities from *R. catesbeiana* pituitaries (Tanaka et al. 1983). This apparent discrepancy is derived, in part, by the use of different bioassays to distinguish LH and FSH biological activities.

Seasonal studies of *R. catesbeiana*, using homologous radioimmunoassays for bullfrog LH and FSH (Licht et al. 1983), demonstrate that gonadotropin concentrations change seasonally with an annual pattern which generally follows the pattern for gonadal steroids. In female bullfrogs, plasma FSH and LH concentrations are low from August to late April, begin to rise in late April or early May, and are highest during the breeding season (Licht et al. 1983). In male Japanese red-bellied newts (*Cynops pyrrhogaster*), the amounts of LH and FSH in the pituitary gland, as determined by bioassays, change seasonally, with levels increasing from September to March and levels decreasing from April to June (Tanaka et al. 1980). Seasonal cycles in gonadotropin concentrations in amphibians, therefore, are correlated with changes in gonadal activity and secondary sexual characteristics.

III. Gonadotropin-Releasing Hormone

A. Structural Heterogeneity

Gonadotropin-releasing hormone [GnRH, used here as a synonym for luteinizing hormone-releasing hormone (LHRH)] is a decapeptide that in mammals has the following amino acid sequence: pGlu-His-Trp-Ser-Tyr-Gly-Leu-Arg-Pro-Gly-NH$_2$ (Schally et al. 1973). The predominant form of GnRH in amphibian brains is indistinguishable from mammalian GnRH, although the GnRHs in birds, reptiles, and teleost fishes differ structurally from mammalian GnRH (King and Millar 1979a). Other studies confirm that a mammalian-like GnRH occurs in the brains of amphibians (Eiden et al. 1982, Sherwood et al. 1986). However, as shown in Figure 2-4, amphibians have other forms of GnRH in brain tissue besides the mammalian-like GnRH. One of these amphibian GnRHs appears to be similar, if not identical, to salmon GnRH (Sherwood, 1986).

B. Neuroanatomical Distribution of GnRH

Immunocytochemical studies of amphibians indicate that GnRH immunoreactivity is localized in neuroanatomical regions similar to those of other vertebrates (reviewed by Barry 1979). In anurans and urodeles, cell bodies with GnRH immunoreactivity occur in the medial septal nucleus and in the anterior preoptic area (Nozaki and Kobayashi 1979, Crim 1985, Jokura and Urano 1986). Figure 2-5 shows the distribution of GnRH immunoreactivity in *Bufo japonicus* and specifically labels the GnRH-immunoreactive perikarya in the anterior preoptic area as being in the diagonal band of Broca (Jokura and Urano 1986). Neuronal fibers containing GnRH immunoreactivity project to the anterior preoptic nucleus, to the median eminence, and to many extrahypothalamic brain areas. Recent studies of *T. granulosa* also have identified GnRH immunoreactivity in the anterior telencephalon in cell bodies and fibers of what is probably the nervus terminalis (Moore et al. 1987), a structure which contains GnRH immunoreactivity in fish (Schreibman et al. 1984) but which had not been known to exist in amphibians. In addition, GnRH has been isolated and purified from sympathetic ganglia and retina of amphibians (Eiden and Eskay 1980; Eiden et al. 1982). This distribution of GnRH immunoreactivity makes it possible for GnRH to be released in or around the preoptic nucleus, third ventricle, median eminence, and many extrahypothalamic areas in the brain as well as in peripheral nerves.

C. Seasonal Changes in GnRH Concentrations

With the development of microdissection procedures, it is possible to quantify GnRH immunoreactivity in specific brain areas from animals in various physiological states. The procedure involves sectioning frozen brain tissue, dissecting out specific brain areas from frozen sections under a microscope, and quantifying GnRH by radioimmunoassay. The microdissection–radioimmunoassay procedure is more powerful than immunocytochemistry for quantifying neuropeptides in specific brain areas, but it is less discriminating than immunocytochemistry for determining neuroanatomical locations of neuropeptides.

Microdissection–radioimmunoassay technique was used to investigate seasonal changes in GnRH immunoreactivity in the infundibulum, rostral hypothalamus, and ventral preoptic area of *T. granulosa* males (Zoeller and Moore 1985). Immunoreactivity of GnRH was found to change seasonally in a pattern that roughly reflects the seasonal cycle of reproduction (Figure 2-6). A dramatic

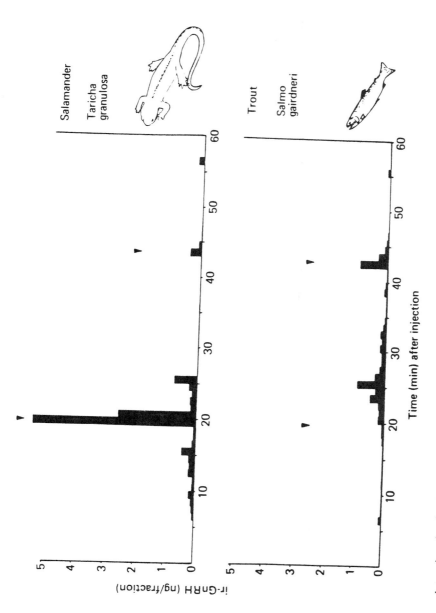

Figure 2-4. Gonadotropin-releasing hormone (GnRH) immunoreactivity in amphibian and fish brain tissues following separation with reverse-phase high-pressure liquid chromatography. Arrows point to the elution times for mammalian GnRH and synthetic salmon GnRH. (From Sherwood et al. 1986.)

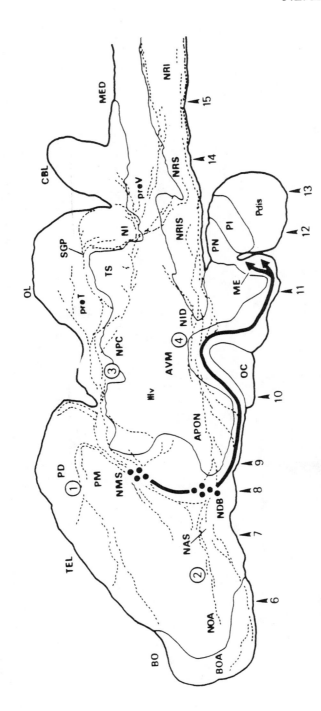

Figure 2-5. Diagrammatic representation of the mid-sagittal plane of the toad (*Bufo japonicus*) brain illustrating the neuroanatomical distribution of immunoreactive gonadotropin-releasing hormone (GnRH) in perikarya (filled circles) and fibers (broken lines). GnRH-containing perikarya occur in the nucleus medialis septi (NMS) and the diagonal band of Broca (NDB) in the anterior preoptic area. The thick line represents the concentration of GnRH-containing fibers that project to the median eminence (ME). (From Jokura and Urano 1986.)

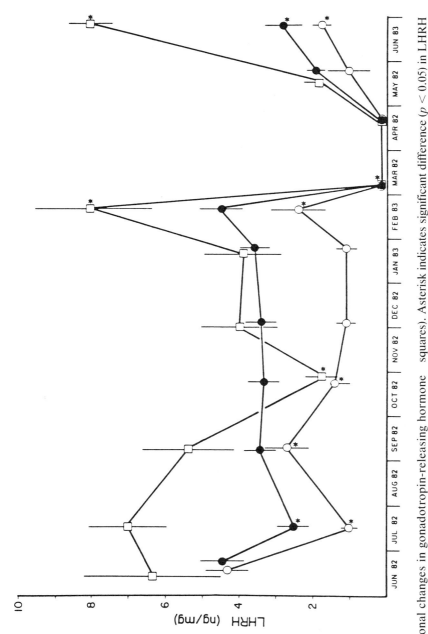

Figure 2-6. Seasonal changes in gonadotropin-releasing hormone (LHRH) concentrations (X ± SE; ng LHRH/mg protein) in three brain areas of *Taricha granulosa*: infundibulum (filled circles), rostral hypothalamus (open circles), and ventral preoptic area (open squares). Asterisk indicates significant difference ($p < 0.05$) in LHRH concentration when compared to the previous mean value for given brain area. (From Zoeller and Moore 1985.)

drop in GnRH immunoreactivity occurs in males collected in March and April, which is when the testes are starting to regress and plasma testosterone concentrations are starting to decrease. The resumption of spermatogenesis and spermiogenesis occurs when relatively high concentrations of GnRH immunoreactivity are in the hypothalamus. Seasonal changes in GnRH immunoreactivity also have been observed in *X. laevis* by radioimmunoassay (King and Millar 1979b) and in *B. japonicus* by immunocytochemistry (Jokura and Urano 1985).

D. Functions of GnRH

The various functions that have been attributed to GnRH in amphibians include (1) stimulation of gonadotropin release from the pituitary gland (Licht 1986, Peter 1986), (2) activation of sexual behaviors in female *X. laevis* and male *T. granulosa* (Kelley 1982, Moore et al. 1982), (3) stimulation of catecholamine release in *Bufo marinus* (Wilson et al. 1984), and (4) stimulation of postsynaptic neuronal activity in *R. catesbeiana* (Jan et al. 1979). Considering this range of functions and the presence of multiple forms of GnRH in amphibians, one intriguing hypothesis is that particular GnRH-like peptides are dedicated to one or more of these purported functions.

The best-documented function for GnRH in amphibians is the stimulation of pituitary gonadotropin secretion. For example, the administration of mammalian GnRH to amphibians can stimulate the release of gonadotropins (Licht 1986, Peter 1986). Given that mammalian-like GnRH is the predominant form in amphibian brain tissue (Sherwood et al. 1986) and that mammalian-like GnRH appears to be localized in the median eminence of the hypothalamus, it seems likely that this molecule functions as the endogenous stimulator of gonadotropin release.

The activation of sexual behaviors by injections of mammalian GnRH into amphibians is consistent with reports that GnRH can activate behavioral responses in other vertebrates (Alderete et al. 1980, Boyd and Moore 1985). In *X. laevis*, systemic injections of mammalian GnRH stimulate sexual receptivity in intact and in ovariectomized females implanted with estradiol plus progesterone, indicating that this behavioral response to GnRH is independent of the pituitary–gonad axis (Kelley 1982). In *T. granulosa* males early in the breeding season, intracerebroventricular (icv) injections of mammalian GnRH activate male clasping behaviors; icv injections of a GnRH antagonist inhibit these behaviors (Moore et al. 1982). The seasonal study of GnRH in *T. granulosa* indicates that hypothalamic GnRH may not be involved in activating male clasping behaviors because there are very low concentrations of GnRH immunoreactivity in the hypothalamus when the males are sexually most active (Zoeller and Moore 1985). Perhaps, GnRH has behavioral functions in extrahypothalamic areas in *T. granulosa*.

There also is evidence that GnRH may function as a neurotransmitter in amphibians (Jan et al. 1979). Neural stimulation of bullfrog sympathetic ganglia induces the release of GnRH-like material from preganglionic elements and causes a late slow excitatory postsynaptic potential (epsp) in ganglionic elements. This epsp response can be blocked by the administration of GnRH antagonists and can be stimulated by the administration of GnRH agonists, supporting the conclusion that GnRH-like molecules in the sympathetic ganglion can function as neurotransmitters. Such a transmitter function for GnRH may be neurophysiologically related to the effects of GnRH on sexual behaviors and catecholamine release.

IV. Arginine Vasotocin in Reproduction

Arginine vasotocin (AVT) is a small neuropeptide that originally was isolated from neurohypophyses of nonmammalian vertebrates and was found to function in the control of hydromineral balance (Acher 1985). Subsequently, AVT has been found to occur in hypothalamic and extrahypothalamic brain regions (Zoeller and Moore 1986a) and to activate reproductive behaviors in many vertebrates (reviewed by Moore 1986).

AVT apparently controls reproductive behaviors in some amphibians. Systemic injections of AVT can induce sexual receptivity in female *Rana pipiens* (Diakow 1978). In *T. granulosa,* icv (but not intraperitoneal) injections of nanogram quantities of AVT can activate male clasping behaviors, suggesting that the site of action for this response is in the brain (Moore and Miller 1983). Furthermore, reproductive behaviors in *T. granulosa* are suppressed by icv injections of an AVP antagonist.

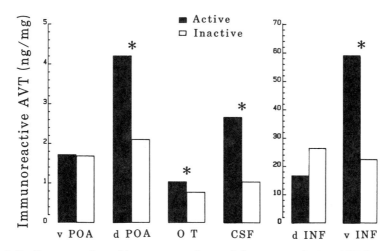

Figure 2-7. Concentration of immunoreactive arginine vasotocin (ng AVT/mg protein) in specific brain areas of *Taricha granulosa* males: ventral and dorsal preoptic areas (vPOA, dPOA), infundibulum (INF), optic tectum (OT), and cerebrospinal fluid (CSF). Asterisk indicates significant difference (p <0.05) between sexually active and inactive males. (Zoeller and Moore 1987.)

High correlation between AVT concentrations in certain brain regions and behavioral state provide evidence that AVT regulates reproductive behavior in *T. granulosa* (Zoeller and Moore 1987). The concentrations of AVT immuno-reactivity in several brain areas are significantly higher in sexually active than in inactive male newts (Figure 2-7). These data, along with the injection studies, support the conclusion that endogenous AVT is causally related to the male's propensity to exhibit male clasping behaviors.

In *T. granulosa* males, the concentrations of AVT immunoreactivity in the optic tectum, but not in the dorsal preoptic area, change seasonally (Zoeller and Moore 1986b). Because the concentrations of AVT immunoreactivity are highest in the optic tectum from February through April, when male newts are most sexually active, these data demonstrate that, in this brain area, AVT is seasonally correlated with the occurrence of sexual behaviors. This study provides evidence that behaviorally active neuropeptides may change seasonally in association with the seasonal cycle of reproduction.

V. Neuroendocrine Activation of Reproduction

A. Acute and Annual Changes in Reproduction

Seasonal cycles of reproduction must be timed so that the individual is prepared to mate or lay eggs when the time is appropriate. Because gametogenesis and morphological development take weeks or months to complete, animals must initiate these developmental changes prior to the breeding season. Amphibians accomplish this with endogenous cycles of reproductive development, cycles that can be accelerated or retarded by particular environmental stimuli (see Section IC). In temperate climates, the predominant environmental stimuli that influence reproductive cycles are temperature and photoperiod, two predictors of seasonal changes in most environments.

Superimposed on the annual cycles in reproduction are short-term, acute, behavioral and neurophysiological changes that regulate exactly when and where reproduction takes place. The time scale for these changes can range from minutes to a few days. Spadefoot toads emerging and mating shortly after heavy rains is an example of acute changes in reproduction. There are numerous other examples where frog chorusing, breeding activity, or migrations to breeding sites are correlated with short-term fluctuations in temperature and moisture (Salthe and Mecham 1974). Such acute changes can be triggered by numerous environmental stimuli, but temperature and moisture are frequently the dominant cues.

B. Activation of Reproductive Functions

Most discussions of reproduction focus on the hormones that activate specific reproductive phenomena, such as the neuropeptides (GnRH), pituitary hormones (LH, FSH, PRL [prolactin]), and gonadal steroid hormones (testosterone and dihydrotestosterone in males; testosterone, progesterone, and estradiol in females). These hormones are only half of the picture; other hormones control reproduction by suppressing or inhibiting particular reproductive events. The balance between these activational and inhibitory mechanisms determines whether an individual will exhibit particular reproductive responses in a given situation.

As in other vertebrates, in amphibians GnRH stimulates the release of gonadotropins, and gonadotropins stimulate gonadal development and steroid hormone secretion. However, amphibians appear to have several GnRH-like molecules (Sherwood et al. 1986), which raises the question of which GnRH-like molecule is the principal regulator of pituitary gonadotropin secretion. The salmonlike GnRH, because it apparently is localized in sympathetic ganglia, probably acts directly on neurons as a neurotransmitter (Sherwood, 1986).

Seasonal studies of GnRH in amphibians (Section IIIC) indicate that changes in hypothalamic GnRH are important to the neuroendocrine regulation of seasonal reproduction. In male *T. granulosa*, hypothalamic GnRH concentrations change seasonally with a pattern that includes a marked decrease in GnRH concentrations when males are most sexually active (Figure 2-6). This decrease in GnRH occurs prior to the regression of testicular tissue. The timing of this decrease in hypothalamic GnRH provides evidence for rejecting two previously proposed hypotheses: (1) that gonadal regression is initiated by a seasonal decrease in gonadal sensitivity to gonadotropins or (2) that gonadal regression is initiated by a seasonal decrease in gonadal sensitivity to gonadotropins. Furthermore, the concentrations of hypothalamic GnRH in *T. granulosa* are elevated from May through August, which is when spermatogenesis and spermiogenesis occur (Specker and Moore 1980, Zoeller and Moore 1985). These seasonal changes in hypothalamic GnRH activity suggest that changes in GnRH secretion play a central role in regulating seasonal reproduction and that changes in pituitary or gonadal sensitivity are less important.

It is not known which hormones cause the seasonal changes in GnRH or, more specifically, which hormones cause the marked decrease in hypothalamic GnRH just prior to testicular regression in *T. granulosa*. One hypothesis is that the decrease in GnRH results from an increased sensitivity to negative feedback by testosterone. Zoeller (1984) conducted experiments to test this hypothesis, but he found that although GnRH appeared to be more sensitive to the negative feedback effects of testosterone in May than in September, his complex results cannot be explained by this simple hypothesis. One complicating factor is an apparent involvement of corticosterone in modulating GnRH secretion in amphibians (see Section VIA).

It also is not known which environmental factors trigger the drop in hypothalamic GnRH content. However, studies with other reproductive responses suggest that the seasonal cycle in GnRH probably would occur in animals held under constant environmental conditions, but that the timing of the cycle probably would be influenced by temperature, photoperiod, and other factors (see Sections IC and VIA).

This discussion has emphasized seasonal changes in GnRH because this chapter focuses on the neuroendocrine control of reproduction. However, there are changes in the pituitary glands and gonads which also may modulate seasonal cycles in reproduction. For example, there is evidence that low ambient temperatures can act directly on gonads to suppress in vitro responses to gonadotropin (Muller 1977). Such responses to low ambient temperatures may mediate the disruption of reproduction that can occur during unseasonably cold weather or during winter hibernation; however, such responses probably are not involved in postbreeding gonadal regression because in temperate zone amphibians, regression occurs when temperatures are increasing, not decreasing. Secondly, there also is evidence that the sensitivity of the gonads to gonadotropin can change seasonally in some amphibians (Lofts 1984). The critical question here is whether gonadal insensitivity to gonadotropin stimulation is the cause or the effect of changes in hypothalamic GnRH. Considering that GnRH concentrations decrease early in the breeding season in *T. granulosa,* it seems most likely that a decrease in GnRH precipitates gonadal change, not vice versa.

C. Activation of Reproductive Behaviors

Reproductive behaviors of amphibians are controlled by gonadal steroid hormones and neuropeptides (see Sections IIC, IIIC, and IV). The gonadal steroids that have been shown to activate reproductive behaviors include estrogen and progesterone in female amphibians (Kelley 1982) and testosterone, dihydrotestosterone, and estrogen in male amphibians (Kelley and Pfaff 1976, Moore 1978, Andreoletti et al. 1983, Moore and Miller 1983, Wetzel and Kelley 1983). Although these studies demonstrate that gonadal steroids can activate reproductive behaviors in some amphibians, numerous other studies have failed to find any effect of gonadal steroid hormones on amphibian reproductive behaviors (reviewed in Moore 1983). These apparently inconsistent results are explained, at least in part, by the fact that gonadal steroids are not the only hormones that regulate amphibian reproductive behaviors.

Gonadal steroids could activate amphibian reproductive behaviors by several different actions. First, gonadal steroids can stimulate neuronal growth and development in birds and mammals (Arnold and Breedlove 1985, Kurz et al. 1986), resulting in sexual dimorphism in behaviorally important neuronal structures. The only evidence of this phenomenon in amphibians comes from a study of *B. japonicus,* which demonstrates the existence of sexual dimorphism in the amygdala and anterior preoptic area (Takami and Urano 1984). Second, gonadal steroids can alter the distribution of hormone receptors in target neurons. For example, in *X. laevis,* implants of estradiol in ovariectomized females can increase the number of progestin binding sites in specific brain areas (Roy et al. 1986), a finding consistent with behavioral studies showing that sexual receptivity in this species is activated by combined treatment with estradiol plus progesterone (Kelley 1982). There also is indirect evidence that gonadal steroids can influence receptors for behaviorally active neuropeptides; in *T. granulosa,* the behavioral responses to AVT injections are abolished by castration and are restored by androgen implants (Moore and Zoeller 1979, Zoeller and Moore 1982). Third, gonadal steroids can influence the synthesis or release of classic neurotransmitters and behaviorally active neuropeptides. For example, the seasonal changes in the concentrations of AVT in the optic tectum of male newts could be the result of seasonal changes in testicular steroids (Zoeller and Moore 1986b). Last, gonadal steroids can influence sexual behaviors by affecting peripheral structures. For example, in *X. laevis*, testosterone can strengthen muscles that are involved in amplectic clasping (Erulkar et al. 1981). In summary, gonadal steroids regulate reproductive behaviors by acting at multiple sites within and outside of the neuroendocrine system.

Amphibian reproductive behaviors have been shown to be influenced by GnRH and AVT (see Sections IIIC and IVA). These behavioral actions are interesting for several reasons. They provide an explanation for the wide neuroanatomical distribution of these neuropeptides. If GnRH only controlled gonadotropin release and AVT only controlled hydromineral balance, then these neuropeptides probably would have a limited distribution in the hypothalamus. Further, GnRH and AVT are likely candidates for mediating short-term behavioral changes; other neuropeptides have been shown to have the capacity to affect target neurons with pronounced and prolonged effects, sometimes lasting for many minutes (Mayeri et al. 1979).

Speculating about the behavioral actions of GnRH and AVT in amphibians, the GnRH in the olfactory region of *T. granulosa* (described in Section IIIB) might modulate neuronal pathways involved in detecting sexual pheromones. Likewise, given the neuroanatomical distribution of AVT, this peptide might act at several sites: in the optic tectum it might modulate neuronal pathways that integrate visual stimuli, in the hindbrain regions it might modulate motor pathways, and in the preoptic nucleus or medial pallium it might modulate central arousal mechanisms. The principal point of these speculations is that GnRH, AVT, and other neuropeptides could control reproductive behaviors by modulating behaviorally important neurons in the sensory pathways, the central nervous system, and the motor pathways.

VI. Neuroendocrine Inhibition of Reproduction

A. Inhibition of Reproductive Hormones

Probably the most dramatic example of short-term inhibition of reproduction occurs when a predator, such as a human, splashes through a breeding pond of frogs or toads—the loud chorusing is replaced by silence. Although the neuroendocrine mediators of this response are unknown, the hormones that mediate other inhibitory effects on reproduction have been identified. Because these inhibitory responses in amphibians have been associated with harsh or noxious environmental conditions, they are collectively referred to here as stress-induced inhibition of reproduction. Stress-induced inhibition of reproduction in amphibians appears to be regulated by the classic stress hormones—corticotropin-releasing hormone (CRH), corticotropin (ACTH), and corticosterone.

Licht and colleagues demonstrated that reproductive hormones are suppressed by the stress of captivity. When bullfrogs are captured and held in collecting sacks or are received from commercial suppliers, the concentrations of gonadal steroids and gonadotropins are low (Licht et al. 1983). Similarly, the suppressive effects of stress on plasma androgen concentrations have been

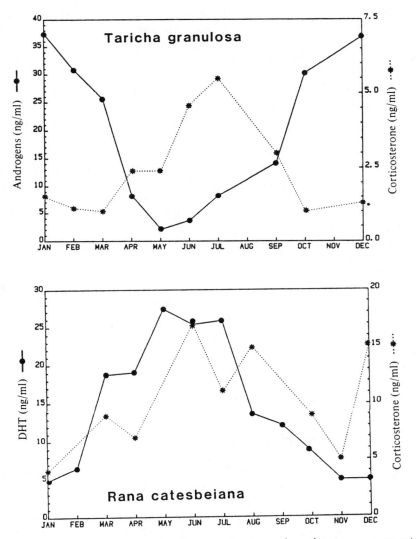

Figure 2-8. Seasonal changes in plasma androgen and corticosterone concentrations in three species of amphibians. Data are from Zoeller and Moore 1985, Licht et al. 1983, D'Istria et al. 1974, Leboulenger et al. 1979.

observed in *T. granulosa* (Moore and Zoeller 1985). To investigate the physiological basis of this stress effect, *T. granulosa* males were exposed to mild stress and then plasma androgen and hypothalamic GnRH concentrations were measured. Because GnRH concentrations were significantly higher and plasma androgen concentrations were significantly lower in stressed males compared to unstressed controls, it appears that stress can inhibit GnRH secretion. In another study, in corticosterone-injected compared to saline-injected males, GnRH concentrations were significantly higher and plasma androgen concentrations were significantly lower. These studies indicate that corticosterone may mediate the stress-induced decrease in androgens by acting at the level of the hypothalamus (directly or indirectly) to inhibit GnRH secretion (Moore and Zoeller 1985).

   T. granulosa males are more sensitive to corticosterone injections during February than September, as measured by changes in GnRH and androgen concentrations (Moore and Zoeller 1985). This apparent seasonal change in sensitivity to corticosterone is interesting because of the reported seasonal cycles of plasma corticosterone concentrations in amphibians (Licht et al. 1983, Pancak and Taylor 1983, Jolivet-Jaudet et al. 1984, Zoeller and Moore 1985). In *R. catesbeiana*, *T. granulosa*, and *R. esculenta* (Fig. 2-8), plasma corticosterone concentrations reach peak levels during the summer, and in the latter two species there is an inverse relationship between plasma corticosterone and androgens. Considering that *T. granulosa* males are most sensitive to corticosterone when corticosterone titers are rising and androgen titers are falling, it is possible that the seasonal rise in plasma corticosterone precipitates the vernal decrease in testicular steroids by blocking GnRH secretion and synthesis (Figures 2-6 and 2-8).

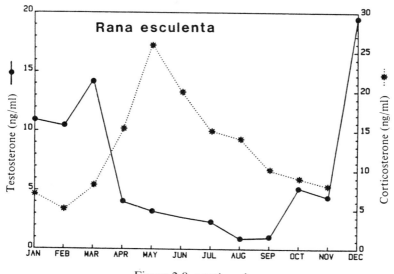

Figure 2-8 *continued.*

B. Inhibition of Reproductive Behaviors

Although *T. granulosa* readily exhibit courtship and mating behaviors under normal laboratory conditions, sexual behaviors sometimes are suppressed in males that are injected repeatedly, handled extensively, or otherwise exposed to harsh conditions. At least four different chemical messengers may be involved (directly or indirectly) in this stress-induced inhibition of sexual behaviors: corticosterone, ACTH, and an opiate peptide (Moore and Miller 1984).

That corticosterone is involved in inhibiting newt sexual behaviors is supported by several types of experiments (Moore and Miller 1984). First, plasma corticosterone concentrations are negatively correlated with the incidence of courtship in *T. granulosa* males. Second, when plasma corticosterone concentrations are experimentally elevated by injections of corticosterone or CRH, sexual behaviors are suppressed in a dose-response manner. Third, when the stress-induced rise in corticosterone is selectively blocked by pretreating males with metyrapone, the negative effects of stress on sexual behaviors are ameliorated. Together, these data support the conclusion that corticosterone is a causal factor in the stress-induced inhibition of sexual behavior in this amphibian.

Because corticosterone secretion is controlled by hypothalamic CRH and pituitary ACTH, it is not surprising that injections of CRH or ACTH can suppress sexual behaviors in *T. granulosa* (Miller and Moore 1983, Moore and Miller 1984). The inhibitory effects of ACTH injections probably result from elevated corticosterone concentrations (Miller and Moore 1983). The inhibitory effect of CRH on sexual behavior appears to be mediated both by corticosterone and some other more direct pathway. Evidence for this more direct pathway comes from several sources, including the findings with *T. granulosa* that CRH injections can stimulate locomotor activities (swimming and walking) independently of ACTH or corticosterone (Moore et al. 1984).

Also, in *T. granulosa*, the stress-induced inhibition of sexual behaviors may involve an endogenous opiate (Deviche and Moore, unpublished). Sexual behaviors of male newts are inhibited by injections of dynorphin or bremazocine (a synthetic kappa receptor agonist). Bremazocine inhibits sexual behaviors in a dose-response manner, and naloxone (a synthetic opioid antagonist) can reverse the effects of bremazocine injections. These data indicate that an opioid peptide that binds to kappa receptors can inhibit sexual behaviors in this amphibian. Presently, more research is needed to determine whether this opiate response is involved in stress-induced inhibition of sexual behaviors.

These recent discoveries that particular hormones can inhibit amphibian sexual behaviors provide further insight into the regulation of short-term changes in behavioral state. They also provide further evidence that the neuroendocrine control of amphibian reproduction involves two general sets of chemical messengers. One set activates reproduction and includes GnRH, LH, FSH, AVT, and gonadal steroids. The other set inhibits reproduction and includes CRH, ACTH, corticosteroids, and opiate peptides.

References

Acher R (1985) Biosynthesis, processing, and evolution of neurohypophysial hormone precursors. In: Kobayashi H, Bern HA, Urano A (eds) Neurosecretion and the Biology of Neuropeptides. Springer-Verlag, New York, pp. 11–25.

Alderete MR, Tokarz RR, Crews D (1980) Luteinizing hormone-releasing hormone and thyrotropin-releasing hormone induction of female sexual receptivity in the lizard, *Anolis carolinensis*. Neuroendocrinology 30:200–205.

Andreoletti GE, Malacarne G, Vellano C (1983) Androgen control of male sex behavior in the crested newt (*Triturus cristatus carnifex* Laur.): castration and sex steroid administration. Horm Behav 17:103–110.

Arnold AP, Breedlove SM (1985) Organizational and activational effects of sex steroids on brain and behavior: a reanalysis. Horm Behav 19:469–498.

Barry J (1979) Immunohistochemistry of luteinizing hormone-releasing hormone-producing neurons of the vertebrates. Int Rev.Cytol 60:179–221.

Bolaffi JL, Lance V, Callard IP, Walsh JM, Idler DR (1979) Identification of 11-keto-testosterone, 11β-hydroxytestosterone, and testosterone in plasma of *Necturus maculosus* (Rafinesque). Gen Comp Endocrinol 38:127–131.

Boyd SK, Moore FL (1985) Luteinizing hormone-releasing hormone facilitates the display of sexual behavior in male voles (*Microtus canicaudus*). Horm Behav 19:252–264.

Brown PS, Brown SC, Specker JL (1984) Osmoregulatory changes during the aquatic-to-terrestrial transition in the rough-skinned newt, *Taricha granulosa*: the roles of temperature and ACTH. Gen Comp Endocrinol 56:130–139.

Callard GV, Petro Z, Ryan KJ (1978) Conversion of androgen to estrogen and other steroids in the vertebrate brain. Am Zool 18:511–523.

Cayrol C, Garnier DH, Deparis P (1985) Comparative plasma levels of androgens and 17β-estradiol in the diploid and triploid newt, *Pleurodeles waltl*. Gen Comp Endocrinol 58:342–346.

Clergue M, Moatti JP, Milhet P (1985) Androgènes et 17β-estradiol plasmatiques de l'urodèle pyrénéen *Euproctus asper* durant la période de reproduction. Gen Comp Endocrinol 58:337–341.

Crews D (1984) Gamete production, sex hormone secretion, and mating behavior uncoupled. Horm Behav 18:22–28.

Crim JW (1985) Immunocytochemistry of luteinizing hormone-releasing hormone and sexual maturation of the frog brain: comparisons of juvenile and adult bullfrogs (*Rana catesbeiana*). Gen Comp Endocrinol 59:424–433.

Diakow C (1978) Hormonal basis for breeding behavior in female frogs: vasotocin inhibits the release call of *Rana pipiens*. Science 199:1456–1457.

D'Istria M, Delrio G, Botte V, Chieffi G (1974) Radioimmunoassay of testosterone, 17β-oestradiol and oestrone in the male and female plasma of *Rana esculenta* during sexual cycle. Steroids Lipids Res. 5:42–48.

Duellman WE, Trueb L (1986) Biology of Amphibians. McGraw-Hill Book Co., New York.

Eiden LE, Eskay RL (1980) Characterization of LRF-like immunoreactivity in the frog sympathetic ganglia: non-identity with LRF decapeptide. Neuropeptides 1:29–37.

Eiden LE, Loumaye E, Sherwood N, Eskay RL (1982) Two chemically and immunologically distinct forms of luteinizing hormone-releasing hormone are differentially expressed in frog neural tissues. Peptides 3:323–327.

Erulkar SD, Kelley DB, Jurman ME, Zemlan FP, Schneider GT, Krieger NR (1981) Modulation of the neural control of the clasp reflex in male *Xenopus laevis* by androgens: a multidisciplinary study. Proc Natl Acad Sci USA 78:5876–5880.

Ewert J-P, Capranica RR, Ingle DJ (1983) Advances in Vertebrate Neuroethology. Plenum Press, New York/London.

Garnier DH (1985) Androgen and estrogen levels in the plasma of *Pleurodeles waltl*, Michah., during the annual cycle. I. Male cycle. Gen Comp Endocrinol 58:376–385.

Houck LD (1977) Life history patterns and reproductive biology. In: Taylor DH, Guttman SI (eds) The Reproductive Biology of Amphibians. Plenum Publishing Corp., New York.

Imai K, Tanaka S, Takikawa H (1985) Annual cycle of gonadotropin and testicular steroid hormones in the Japanese red-bellied newt. In: Loft B, Holmes WN (eds) Current Trends in Comparative Endocrinology. Hong Kong University Press, Hong Kong, pp 247–249.

Ingle D, Crews D (1985) Vertebrate neuroethology: definitions and paradigms. Ann Rev Neurosci 8:457–494.

Jan YN, Jan LY, Kuffler SW (1979) A peptide as a possible transmitter in sympathetic ganglia of the frog. Proc Natl Acad Sci USA 76:1501–1505.

Jokura Y, Urano A (1985) An immunohistochemical study of seasonal changes in luteinizing hormone-releasing hormone and vasotocin in the forebrain and the neurohypophysis of the toad, *Bufo japonicus* Gen Comp Endocrinol 59:238–245.

Jokura Y, Urano A (1986) Extrahypothalamic projection of luteinizing hormone-releasing hormone fibers in the brain of the toad, *Bufo japonicus*. Gen Comp Endocrinol 62:80–88.

Jolivet-Jaudet G, Inoue M, Takada K, Ishii S (1984) Circannual changes in plasma aldosterone levels in *Bufo japonicus formosus*. Gen Comp Endocrinol 53:163–167.

Jørgensen CB, Hede K-E, Larsen LO (1978) Environmental control of annual ovarian cycle in the toad *Bufo bufo bufo* L.: role of temperature. In: Assenmacher I, Farner DS (eds) Environmental Endocrinology. Springer-Verlag, New York, pp. 28–36.

Kelley DB (1980) Auditory and vocal nuclei in the frog brain concentrate sex hormones. Science 207:553–555.

Kelley DB (1982) Female sex behaviors in the South African clawed frog, *Xenopus laevis*: gonadotropin-releasing, gonadotropic, and steroid hormones. Horm Behav 16:158–174.

Kelley DB, Pfaff DW (1976) Hormone effects on male sex behavior in adult South African clawed frogs, *Xenopus laevis*. Horm Behav 7:159–182.

Kelley DB, Pfaff DW (1978) Generalizations from comparative studies on neuroanatomical and endocrine mechanisms of sexual behavior. In: Hutchison JB (ed) Biological Determinants of Sexual Behavior. John Wiley and Sons, Chichester.

King JA, Millar RP (1979a) Heterogeneity of vertebrate luteinizing hormone-releasing hormone. Science 206:67–69.

King JA, Millar RP (1979b) Hypothalamic luteinizing hormone-releasing hormone content in relation to the seasonal reproductive cycle of *Xenopus laevis*. Gen Comp Endocrinol 39:309–312.

Kurz EM, Sengelaub DR, Arnold AP (1986) Androgens regulate the dendritic length of mammalian motoneurons in adulthood. Science 232:395–398.

Leboulenger F, Delarue C, Tonon MC, Jegou S, Leroux P, Vaudry H (1979) Seasonal study of interrenal function of the European green frog, *in vivo* and *in vitro*. Gen Comp Endocrinol 39:388–396.

Lecouteux A, Garnier DH, Bassez T, Joly J (1985) Seasonal variations of androgens, estrogens, and progesterone in the different lobules of the testis and the plasma of *Salamandra salamandra*. Gen Comp Endocrinol 58:221–221.

Licht P (1979) Reproductive endocrinology of reptiles and amphibians: gonadotropins. Ann Rev Physiol 41:337–351.

Licht P (1986) Suitability of the mammalian model in comparative reproductive endocrinology. In: Ralph CL (ed) Comparative Endocrinology: Developments and Directions. Alan R. Liss, Inc., New York, pp. 95–114.

Licht P, McCreery BR, Barnes R, Pang R (1983) Seasonal and stress related changes in plasma gonadotropins, sex steroids, and corticosterone in the bullfrog, *Rana catesbeiana*. Gen Comp Endocrinol 50:124–145.

Lofts B (1984) Amphibians. In: Lamming GE (ed) Marshall's Physiology of Reproduction, Vol. 1. Reproductive Cycles of Vertebrates. Churchill Livingstone, New York.

Mayeri E, Brownell P, Branton WD, Simon SB (1979) Multiple, prolonged actions of neuroendocrine bag cells on neurons in *Aplysia*. I. Effects on bursting pacemaker neurons. J Neurophysiol 42:1165–1185.

Mendonca MT, Licht P, Ryan MJ, Barnes R (1985) Changes in hormone levels in relation to breeding behavior in male bullfrogs (*Rana catesbeiana*) at the individual and population levels. Gen Comp Endocrinol 58:270–279.

Miller LJ, Moore FL (1983) Intracranial administration of corticotropin-like peptides increases incidence of amphibian reproductive behavior. Peptides 4:729–733.

Moore FL (1978) Differential effects of testosterone plus dihydrotestosterone on male courtship of castrated newts, *Taricha granulosa*. Horm Behav 11:202–208.

Moore FL (1983) Behavioral endocrinology of amphibian reproduction. BioScience 33:557–561.

Moore FL (1986) Behavioral actions of neurohypophysial peptides. In: Crews D (ed) Psychobiology of Reproductive Behavior: An Evolutionary Perspective. Prentice Hall, Englewood Cliffs, NJ, pp. 61–87.

Moore FL, Miller LJ (1983) Arginine vasotocin induces sexual behavior of newts by acting on cells in the brain. Peptides 4:97–102.

Moore FL, Miller LJ (1984) Stress-induced inhibition of sexual behavior: corticosterone inhibits courtship behaviors of a male amphibian (*Taricha granulosa*). Horm Behav 18:400–410.

Moore FL, Muller CH (1977) Androgens and male mating behavior in rough-skinned newts, *Taricha granulosa*. Horm Behav 9:309–320.

Moore FL, Zoeller RT (1979) Endocrine control of amphibian sexual behavior: evidence for a neurohormone-androgen interaction. Horm Behav 13:207–213.

Moore FL, Zoeller RT (1985) Stress-induced inhibition of reproduction: evidence of suppressed secretion of LH-RH in an amphibian. Gen Comp Endocrinol 60:252–258.

Moore FL, McCormack C, Swanson L (1979a) Induced ovulation: effects of sexual behavior and insemination on ovulation and progesterone levels in *Taricha granulosa*. Gen Comp Endocrinol 39:262–269.

Moore FL, Muller CH, Specker JL (1979b) Origin and regulation of plasma dihydrotestosterone and testosterone in the rough-skinned newt, *Taricha granulosa*. Gen Comp Endocrinol 38:451–456.

Moore FL, Muske L, Propper CR (1987) Regulation of reproductive behaviors in amphibians by LHRH. Edited by by M Schwanzel-Fukuda and LS Demski. *Annals of the New York Academy of Science* (in press).

Moore FL, Miller LJ, Spielvogel SP, Kubiak T, Folkers K (1982) Luteinizing hormone-releasing hormone involvement in the reproductive behavior of a male amphibian. Neuroendocrinology 35:212–216.

Moore FL, Spielvogel SP, Zoeller RT, Wingfield J (1983) Testosterone-binding protein in a seasonally breeding amphibian. Gen Comp Endocrinol 49:15–21.

Moore FL, Roberts J, Bevers J (1984) Corticotropin-releasing factor (CRF) stimulates locomotor activity in intact and hypophysectomized newts (Amphibia). J Exp Zool 231:331–333.

Muller CH (1977) *In vitro* stimulation of 5α-dihydrotestosterone and testosterone secretion from bullfrog testis by nonmammalian and mammalian gonadotropins. Gen Comp Endocrinol 33:109–121.

Norris DO, Norman MF, Pancak MK, Duvall D (1985) Seasonal variations in spermatogenesis, testicular weights, vasa deferentia, and androgen levels in neotenic male tiger salamanders, *Ambystoma tigrinum*. Gen Comp Endocrinol 60:51–57.

Nozaki M, Kobayashi H (1979) Distribution of LHRH-like substance in the vertebrate brain as revealed by immunohistochemistry. Arch Histol Jpn 42:201–219.

Pancak MK, Taylor DH (1983) Seasonal and daily plasma corticosterone rhythms in American toads, *Bufo americanus*. Gen Comp Endocrinol 50:490–497.

Peter RE (1986) Structure-activity studies on gonadotropin-releasing hormone in teleosts, amphibians, reptiles and mammals. In: Ralph CL (ed) Comparative Endocrinology: Developments and Directions. Alan R. Liss, Inc., New York, pp. 75–94.

Pierantoni R, Iela L, Delrio G, Rastogi RK (1984) Seasonal plasma sex steroid levels in the female *Rana esculenta*. Gen Comp Endocrinol 53:126–134.

Polzonetti-Magni A, Botte V, Bellini-Cardellini L, Gobbetti A, Crasto A (1984) Plasma sex hormones and post-reproductive period in the green frog, *Rana esculenta* complex. Gen Comp Endocrinol 54:372–377.

Rastogi RK, Iela L, Delrio G, Di Meglio M, Russo A, Chieffi G (1978) Environmental influence on testicular activity in the green frog, *Rana esculenta*. J Exp Zool 206:49–64.

Roy EJ, Wilson MA, Kelley DB (1986) Estrogen-induced progestin receptors in the brain and pituitary of the South African clawed frog, *Xenopus laevis*. Neuroendocrinology 42:51–56.

Salthe SN, Mecham JS (1974) Reproductive and courtship patterns. In: Lofts B (ed) Physiology of the Amphibia, Vol. II. Academic Press, New York, pp. 309–521.

Santa-Coloma TA, Fernandez S, Charreau EH (1985) Characterization of a sexual steroid binding protein in *Bufo arenarum*. Gen Comp Endocrinol 60:273–279.

Schally AV, Arimura A, Kastin AJ (1973) Hypothalamic regulatory hormones. Science 179:341–344.

Schmidt RS (1980) Development of anuran calling circuits: effects of testosterone propionate injections. Gen Comp Endocrinol 41:80–83.

Schreibman MP, Margolis-Kazan H, Halpern-Sebold L (1984) Structural and functional links between olfactory and reproductive systems: puberty-related changes in olfactory epithelium. Brain Res 302:180–183.

Sherwood NM (1986) Evolution of a neuropeptide family: gonadotropin-releasing hormone. Am Zool 26:1041–1054.

Sherwood NM, Zoeller RT, Moore FL (1986) Multiple forms of gonadotropin-releasing hormone in amphibian brains. Gen Comp Endocrinol 61:313–322.

Siboulet R (1981) Variations saisonnières de la teneur plasmatique en testostérone et dihydrotestostérone chez le crapaud de Mauritanie (*Bufo mauritanicus*). Gen Comp Endocrinol 43:71–75.

Specker JL, Moore FL (1980) Annual cycle of plasma androgens and testicular composition in the rough-skinned newt, *Taricha granulosa*. Gen Comp Endocrinol 42:297–303.

Takami S, Urano A (1984) The volume of the toad medial amygdala-anterior preoptic complex is sexually dimorphic and seasonally variable. Neurosci Lett 44:253–258.

Tanaka S, Iwasawa H (1979) Annual change in testicular structure and sexual character of the Japanese red-bellied newt, *Cynops pyrrhogaster*. Zool Mag 88:295–303.

Tanaka S, Hanaoka Y, Takikawa H (1980) Seasonal changes in gonadotropin activity of the pituitary gland in the adult male Japanese red-bellied newt, *Cynops pyrrhogaster pyrrhogaster*. Zool Mag 89:187–191.

Tanaka S, Hanaoka Y, Wakabayashi K (1983) A homologous radioimmunoassay for bullfrog basic gonadotropin. Endocrinol Jpn 30:71–78.

Wada M, Gorbman A (1977) Relation of mode of administration of testosterone to evocation of male sex behavior in frogs. Horm Behav 8:310–319.

Werner JK (1969) Temperature-photoperiod effects on spermatogenesis in the salamander *Plethodon cinereus*. Copeia 1969:592–601.

Wetzel DM, Kelley DB (1983) Androgen and gonadotropin effects on male mate calls in South African clawed frogs, *Xenopus laevis*. Horm Behav 17:388–404.

Wilson JX, Van Vliet BN, West NH (1984) Gonadotropin-releasing hormone increases plasma catecholamines and blood pressure in toads. Neuroendocrinology 39:437–441.

Zoeller RT (1984) The role of luteinizing hormone-releasing hormone in the neuroendocrine control of seasonal reproduction in male rough-skinned newts, *Taricha granulosa*. Ph.D. thesis, Oregon State University, Corvallis.

Zoeller RT, Moore FL (1982) Duration of androgen treatment modifies behavioral response to arginine vasotocin in *Taricha granulosa*. Horm Behav 16:23–30.

Zoeller RT, Moore FL (1985) Seasonal changes in luteinizing hormone-releasing hormone concentrations in microdissected brain regions of male rough-skinned newts (*Taricha granulosa*). Gen Comp Endocrinol 58:222–230.

Zoeller RT, Moore FL (1986a) Arginine vasotocin (AVT) immunoreactivity in hypothalamic and extrahypothalamic areas of an amphibian brain. Neuroendocrinology 42:120–123.

Zoeller RT, Moore FL (1986b) Correlation between immunoreactive vasotocin in optic tectum and seasonal changes in reproductive behaviors of male rough-skinned newts. Horm Behav 20:148–154.

Zoeller RT, Moore FL (1987) Brain arginine vasotocin concentrations related to sexual behaviors and hydromineral balance in an amphibian. Horm Behav (in press).

Chapter 3

Circadian Organization in Lizards: Perception, Translation, and Transduction of Photic and Thermal Information

HERBERT UNDERWOOD

I. Introduction

All organisms show significant daily fluctuations in a host of biochemical, physiological, and behavioral parameters. The term "circadian" (*circa*, about; *dies*, a day) has been applied to those daily rhythms which will persist under constant conditions. Circadian rhythms, therefore, are overt expressions of an internal biological clock. Circadian rhythms are ubiquitous among eucaryotic organisms and undoubtedly confer significant selective advantages since they allow an organism not only to coordinate internal events but to coordinate internal events with events in the external world. It has been speculated that the selection pressure which led to the evolution of circadian rhythmicity was the daily shower of radiation from the sun (Paietta 1982). Biochemical events which were particularly sensitive to the sun's ultraviolet irradiation were confined to the dark phase of the daily light–dark cycle. Consequently, a partitioning of the cell's metabolism occurred with at least some biochemical pathways operating during the night and some during the day. With the evolution of multicellular organisms, the control of the organism's multiple rhythms became more centralized. Among vertebrates, for example, the pineal organ and the suprachiasmatic nuclei of the hypothalamus have been identified as major circadian pacemakers. Concomitant with the evolution of discrete centralized pacemakers, the coordination and phasing of the myriad overt circadian rhythms in different cells and tissues became subserved by the two classical communication systems; that is, nerves and hormones. The coupling of the clock to the photic environment has undoubtedly also undergone a change from a direct photosensitivity toward the involvement of more specialized photoreceptors, such as the eyes. Interestingly, however, a "direct" photosensitivity may have been retained by at least some of the pacemaking areas involved in circadian organization (i.e., submammalian pineal).

This chapter focuses on four important questions about the perception, translation, and transduction of circadian information in a pivotal class of vertebrates, the reptiles, and emphasizes the important role of the pineal organ. First, what is the role of the pineal organ within the lizard's circadian system? Second, how is the pineal coupled to the rest of the circadian system? Third, how do environmental stimuli (light and temperature) control pineal function? Fourth, what is the nature of the photoreceptive inputs into the lizard's circadian system?

II. Pineal Organs: Evolution of Structure and Function

Pineal organs are virtually ubiquitous in vertebrates (Quay 1979, Hamasaki and Eder 1977). Embryologically, the pineal originates as an evagination of the roof of the diencephalon. Whereas birds and mammals possess only a pineal organ, lower vertebrates may, depending on species, possess a second component which originates either in the same vicinity as the pineal or as an actual outpouching of the pineal itself. This second component (called a parapineal organ in general) may remain connected to the brain via a stalk or actually detach from the brain and occupy a position beneath the skull. In lizards this component is referred to as a "parietal eye" and, as its name implies, is remarkably eyelike in morphology, possessing a well-defined retina, lens, and cornea (Quay 1979, Hamasaki and Eder 1977). The parietal eye possesses ganglion cells whose axons can project to the brain, and in some cases the parietal eye nerve may also innervate the pineal organ.

The principal cell type of the pineal organs of lower vertebrates are cells which have some characteristics of photosensory cells (Collin 1979, Collin and Oksche 1981). For example, these cells may have outer segments comprised of stacks or whorls of disks although the outer segments are typically more disorganized or rudimentary in appearance than the outer segments of classical rods or cones. The principal cell type of the pineals of birds and mammals is the pineal parenchymal cell or pinealocyte, which is believed to be derived from the photosensory cells of lower vertebrates (Collin 1979, Collin and Oksche 1981, Juillard et al. 1983). The pinealocyte is secretory in appearance but, in some avian species, very rudimentary outer segments can be seen. Concomitant with these changes in the appearance of the pineal cell type from lower to higher vertebrates, there is a shift in the nature of the pineal innervation; that is, the amount of pineal-petal (efferent) innervation increases while the amount of pinealo-fugal (afferent) innervation decreases.

A photosensory role of the pineal (and parapineal) organs of fish, amphibians, and lizards is not only supported by the appearance of these organs but also by neurophysiological investigations (Quay 1979, Hamasaki and Eder 1977). In lizards the neural activity of the parietal eye shows wavelength-dependent (chromatic) responses to illumination while the pineal tends to show achromatic responses to illumination. Despite the photosensory nature of the pineals of lower vertebrates, they are also capable of considerable biochemical activity. For example, there is evidence that the synthesis of indoleamines occurs in the photosensory-type cells of lower vertebrates (Collin 1979, Collin and Oksche 1981).

III. Pineal Organs: Role in Circadian Organization

Activity rhythms are commonly used as an assay of the state of an animal's circadian clock because the circadian activity rhythm is relatively easy to assay and requires no restraints upon the animal. Pinealectomy has major effects on the circadian activity rhythm of lizards (Underwood 1977, 1981, 1983a,b, 1984). Figure 3-1 shows examples of the effects of pinealectomy on the activity rhythms of three species of iguanid lizards (*Sceloporus occidentalis, S. olivaceus, Anolis carolinensis*). The pinealectomies were performed on the lizards while they were held under conditions of constant temperature and illumination. Under constant conditions circadian rhythms are said to be "freerunning"; that is, they express their endogenous circadian periodicity (the period of the rhythm is the time between recurring phases of the rhythm on subsequent cycles, such as the time between successive activity onsets). Before pinealectomy all of the lizards depicted in Figure 3-1 showed a well-defined freerunning rhythm. After pinealectomy, however, the rhythm showed either (1) a splitting of the single circadian activity rhythm into two circadian components which freeran with markedly different periods (Figure 3-1A), (2) a change in the period of the freerunning rhythm (Figure 3-1B), or (3) arrhythmicity (continuous activity with no discernible activity–rest rhythm) (Figure 3-1C). The kind of effect observed after pinealectomy is only partly species specific. In *S. olivaceus* about 50% of pinealectomized lizards exhibit splitting or period changes in constant conditions and 50% show arrhythmicity (Underwood 1977). In *S. occidentalis* the majority of pinealectomized lizards show period changes and less than 20% show arrhythmicity, whereas virtually all pinealectomized *A. carolinensis* are arrhythmic under constant conditions (Underwood 1981, 1983a). The persistence of circadian rhythmicity after pinealectomy and the splitting into two components clearly shows that the pineal is not the only site involved in generating circadian rhythmicity; if it were, pinealectomy should always generate arrhythmicity. The existence of more than one circadian clock within pinealectomized lizards is supported by two observations. First, the splitting behavior seen in Figure 3-1A clearly shows that at least two circadian pacemakers remain. Second, the activity pattern of the lizard *A. carolinensis* suggests that at least two circadian pacemakers remain after pinealectomy (Underwood 1983a). For example, if anoles entrained to 12L:12D are pinealectomized (Figure 3-1C) the activity pattern begins showing two different phase relationships to the LD cycle. The results from *Anolis* are consistent with the hypothesis that pinealectomy alters the phase relationships of two (or more) circadian oscillators with the LD cycle. Recently, Menaker and Wisner (1983) have shown a circadian rhythm in melatonin secretion from the *Anolis* pineal in vitro showing that the *Anolis* pineal is, itself, the locus of a circadian clock (or clocks). Clearly, therefore, the lizard circadian system is *multioscillator* in nature. At first glance, the fact that pinealectomy can have different kinds of effects, even within a given species, may seem puzzling. However, this behavior is understandable in terms of the multioscillator structure of the lizard's circadian system. Clearly a multioscillator system must have structure insofar as (1) the oscillators comprising the system must normally exhibit the same periodicity and (2) the oscillators must bear fixed phase relationships to one another. The maintenance of a common periodicity and fixed phase relationships can be accomplished either via mutual coupling among the circadian pacemakers and/or the participation of a master circadian pacemaker which would control both the period and phase of its subordinate pacemakers.

This kind of a multioscillator system can explain the several different effects of pinealectomy if it is assumed that the pineal acts either as a coupling mechanism among the various circadian oscillators or as a master circadian pacemaker. Of these two possibilities, it seems more likely that the pineal's role is that of a pacemaker rather than a coupler since the pineal organ of the lizard *A. carolinensis* can exhibit circadian rhythmicity in vitro (Menaker and Wisner 1983). After removal of a pacemaker (the pineal), the remaining oscillators will either become uncoupled, in which case splitting or arrhythmicity will be seen, or the oscillators may retain sufficient coupling to prevent dissociation in which case a change in the period of the coupled system will likely occur (cf. Figure 3-1).

The effects of pinealectomy on the response of the activity rhythm to single light pulses presented to freerunning lizards (*S. occidentalis*) is also compatible with the idea that the pineal acts to phase and coordinate subordinate oscillators (Figure 3-2A) (Underwood 1983b). Single pulses of an entraining agent, such as light, applied at different phases of an otherwise freerunning activity rhythm can elicit phase shifts in the rhythm whose magnitude and direction are a function of the phase at which the pulse was administered (Aschoff 1965, Pittendrigh 1965). A graph of these phase shifts as a function of the phase of the activity rhythm at which the pulse was administered yields the phase response curve (PRC). The PRC can be used to describe how entrainment occurs and also gives insight into the organization of the circadian system. Any entraining stimulus must act to "reset" the period of the endogenous circadian oscillation to match the period of the entraining stimulus. Pittendrigh (1965), for example, has developed a model to describe the mechanism of entrainment in which the phase shift ($\Delta\phi$) that must occur each cycle to induce entrainment must equal the difference between the endogenous period of the circadian clock (τ) and the period of the entraining stimulus (T); that is, $\Delta\phi = \tau - T$. The stimulus, therefore, not only controls the *period* of the rhythm but also its *phase* and the phase relationship between the stimulus and the rhythm can be predicted by reference to a PRC. The distortion that pinealectomy causes in the PRC to light, relative to the PRC for intact (nonpinealectomized) lizards (Figure 3-2A),

---

Figure 3-1. The effect of removing the pineal organ on the circadian activity rhythm in three lizard species: (A) *Sceloporus olivaceus*, (B) *S. occidentalis*, (C) *Anolis carolinensis*. In (A) and (B) the lizards were held in LL. In (C) the lizard was pinealectomized before the beginning of the record, placed on 12L:12D, and finally placed into DD. To aid in interpretation, the activity records of (A) and (B) are presented in duplicate, the right half of each displaced one day above the left; each horizontal line, therefore, represents 48 hours of recording. Deflections of the baseline represent activity. (Figure 3-1 continues on pages 52–53.)

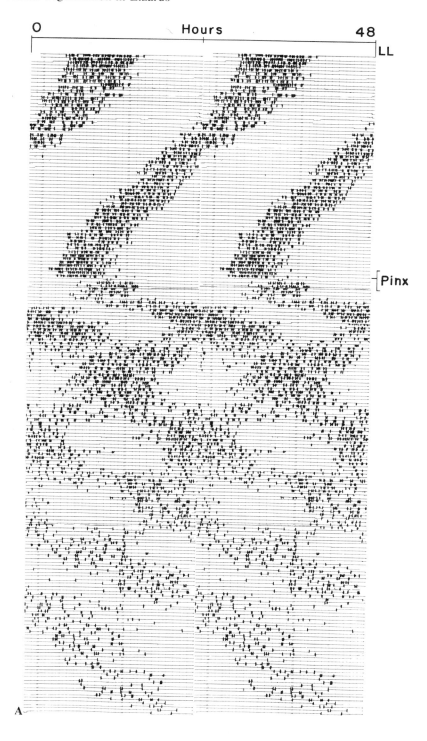

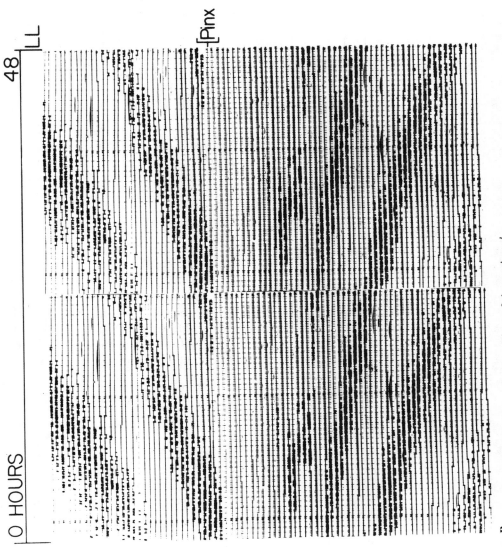

Figure 3-1 continued.

C   Figure 3-1 *continued.*

would be expected if the pineal acts as a pacemaker within a multioscillator circadian system (Underwood 1983b).

Currently, a considerable body of evidence supports the hypothesis that the rhythmic production and secretion of a pineal indoleamine—melatonin—is the mechanism by which the pineal is coupled to the rest of the system.

IV. Melatonin's Role Within the Lizard's Circadian System

Melatonin (*N*-acetyl-5-methoxytryptamine) is one of several methoxyindoles produced in the pineal organ (Reiter 1982). The pathway for melatonin synthesis is as follows: tryptophan → 5-hydroxytryptophan → serotonin $\xrightarrow{(1)}$ *N*-acetyl-serotonin $\xrightarrow{(2)}$ melatonin. Step (1) is catalyzed by the enzyme *N*-acetyltransferase (NAT) and (2) is catalyzed by hydroxyindole-*O*-methyltransferase (HIOMT). Melatonin levels at night are higher than daytime levels regardless of the habits

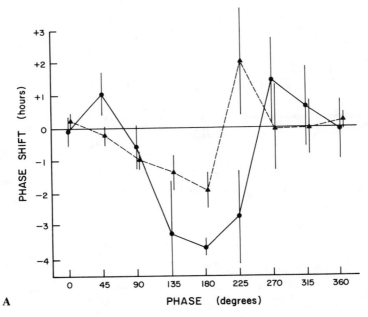

Figure 3-2. Phase response curves to light (A and C) or to melatonin (B) in *S. occidentalis*. The plots show average phase shifts (±SE) measured in hours as a function of the circadian phase at which the stimulus (light/melatonin) was applied in degrees (0 phase marks activity onset.) (A) Response of pinealectomized (solid line) or control (dashed line) lizards to 6-h light pulses. (B) Response to single (10 μg) injections of melatonin (solid line) or to saline (dashed line). (C) Response to 6-h light pulses in control lizards (dashed line) or in lizards previously subjected to bilateral optic nerve section (solid line).

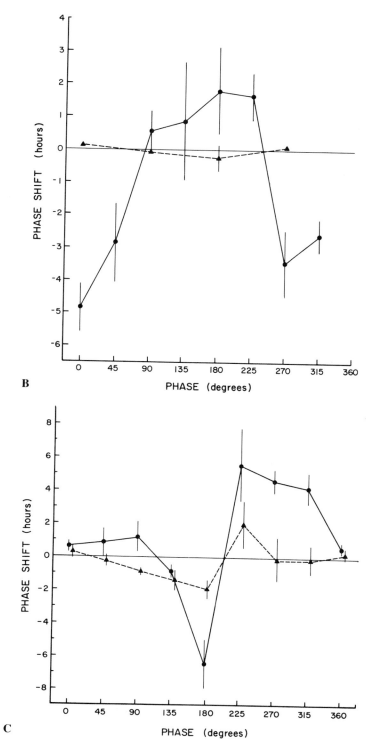

B

C

Figure 3-2 *continued*.

(nocturnal or diurnal) of the animal. It is widely held that the rhythm of melatonin synthesis is due to a rhythm in activity of NAT since HIOMT levels tend to be relatively constant throughout the day. Interest has focused on melatonin for several reasons. (1) Historically, melatonin was the first methoxyindole isolated and it had the interesting property of causing contraction of dermal melanophores in lower vertebrates (Lerner et al. 1958). (2) HIOMT activity was thought to be uniquely localized in pineal tissue making melatonin, and related methoxyindoles, unique pineal products. (3) Melatonin is not stored in pineals but is rapidly secreted into the blood. Accordingly, melatonin could act as a unique pineal hormone. Recently it has been shown that, in some species, melatonin can be synthesized at a few extrapineal sites such as the eyes and Harderian glands (Ralph 1981). However, in those cases examined to date the pineal seems to be the main, if not the exclusive, source of blood-borne melatonin (Ralph 1981, Underwood et al. 1984).

A role for melatonin within the lizard circadian system is supported by several kinds of evidence. Exogenous melatonin administration can cause changes in the freerunning period when given continuously; it can entrain the circadian system when given as a daily injection; and finally, a PRC to melatonin exists (Underwood 1979, 1981, 1986a, Underwood and Harless 1985).

Subcutaneous silastic implants which release low levels of melatonin (10 μg/day) for long periods elicit a lengthening in the freerunning activity rhythms of the lizards *S. olivaceus* and *S. occidentalis* (Figure 3-3) (Underwood 1979). In a few cases this level of melatonin causes arrhythmicity. Silastic melatonin implants also caused a lengthening in τ in pinealectomized and pinealectomized–blinded (by enucleation) *S. occidentalis*, showing that the exogenous melatonin is not acting via the pineal or eyes but is presumably acting at a more central level (Underwood 1981). Interestingly, however, the lengthening effect of melatonin on τ is significantly greater in pinealectomized lizards than in intact ones (Underwood 1981).

Every-other-day injections of melatonin (10 μg), administered at the same time of day, can entrain the activity rhythm of *S. occidentalis* to a period of 24 h (Figure 3-4) (Underwood and Harless 1985). In most cases activity onsets of this diurnal lizard followed the time of injection by 11 h (range 6.5–13 h). This kind of phase relationship between the time of melatonin administration and activity onset would be expected if, in fact, the exogenous melatonin pulse is "mimicking" the entraining effects of endogenous melatonin production, which normally peaks at night. Although these lizards retained their pineals, any endogenous melatonin rhythms were likely to have been masked by the high exogenous levels administered. In two cases, however, activity onsets occurred only 2 h after the time of the melatonin injections (Underwood and Harless 1985). These data suggest that two different phase relationships can occur between the time of melatonin administration and activity onsets but one phase (~165°) is preferred. Melatonin can have a tranquilizing or sleep-inducing effect on some animals (Lieberman 1985). Accordingly, it could be hypothesized that melatonin's effectiveness as an entraining agent is an indirect one; that is, melatonin induces sleep which in turn entrains the system. According to this

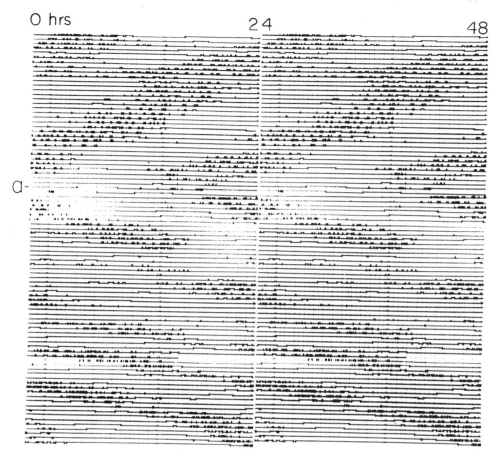

Figure 3-3. Activity record of a lizard (*S. occidentalis*) freerunning in dim LL in which a silastic capsule containing melatonin was implanted on the day marked "a." Melatonin elicited a significant lengthening in the freerunning period.

hypothesis, therefore, any sleep-inducing agent could potentially act as an entraining stimulus. However, the fact that activity onsets immediately followed the time of melatonin administration in two *S. occidentalis* argues against the sleep-induction hypothesis and is more compatible with the idea that the melatonin itself directly entrains the circadian system.

The fact that every-other-day injections of melatonin can entrain the activity rhythm of *S. occidentalis*, with a relatively consistent phase relationship, implies that a PRC to melatonin must exist. A PRC was generated by injecting freerunning *S. occidentalis* with melatonin (10 μg) once every 14 days (Figure 3-2B) (Underwood 1986a). The magnitude and direction of the melatonin-induced phase shift was plotted versus the phase of the activity rhythm at which the melatonin was injected (Figure 3-2B). The PRC shows that injections admin-

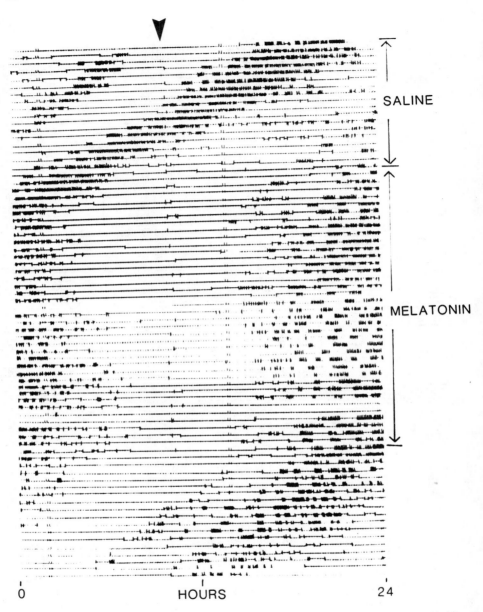

Figure 3-4. Activity record of a lizard (*S. occidentalis*) exposed to dim LL and injected with saline and then melatonin. The time of day of the injection (administered every other day) is indicated by the vertical arrow at the top of the record.

istered between mid-subjective day and early subjective night (6 to 15 h after activity onset) elicit phase advances in the activity rhythm whereas injections given at other phases of the activity cycle induce phase delays.

In summary, exogenously administered melatonin has several effects on the lizard's circadian system—it can affect freerunning rhythms, it can act as an entraining stimulus, and it can induce phase-dependent shifts in the activity rhythm. Also, removal of the pineal, which is a primary melatonin producing site in lizards, can disrupt circadian organization. Furthermore, the circadian system in lizards is multioscillator in nature with a circadian pacemaker(s) residing within the pineal and two (or more) residing elsewhere. Taken together, these data are consistent with a model in which a circadian pacemaker(s) in the pineal drives a daily rhythm in melatonin synthesis and secretion which, in turn, entrains circadian oscillators located elsewhere. Obviously, in addition to discerning how the pineal might control the rest of the circadian system, it is also important to describe how cycles in the physical environment, namely, light and temperature, can act to entrain the pineal melatonin rhythm.

V. Entrainment: Light and Temperature

The daily light–dark (LD) cycle is a universal entraining stimulus for all organisms. In addition, among the vertebrates, daily temperature cycles are also effective entraining agents for poikilotherms. Although other entraining agents are known (i.e., noise, barometric pressure) for at least some species, they play a minor role relative to light and temperature in poikilotherms.

Twenty-four-hour LD cycles can readily entrain the pineal melatonin rhythm (PMR) in lizards (Underwood 1985a). Figure 3-5A shows the PMR of the lizard *A. carolinensis* exposed to either 18L:6D or 6L:18D and constant temperature conditions. The length of the photoperiod clearly affects several features of the PMR including the phase relationship between the melatonin rhythm (i.e., peak melatonin levels) and the LD cycle as well as the amplitude and duration of the PMR. Figure 3-5A also shows that the melatonin levels in the eyes of anoles are either very low or nondetectable and are not rhythmic. Melatonin levels in the Harderian glands of *Anolis* are also low and nonrhythmic (Underwood, unpublished).

Twenty-four-hour temperature cycles are also extremely effective entraining stimuli for the pineal melatonin rhythm of *Anolis* (Underwood 1985a). For example, when anoles are exposed for several weeks to a daily temperature cycle (i.e., 12 h at 20°C and 12 h at 32°C) under constant lighting conditions, the PMR peaks in the cool (20°C) phase of the temperature cycle (Figure 3-5B) (Underwood 1985a). The fact that the rhythm is phase-locked to the cool phase of the temperature cycle is compatible with the thermal ecology of this animal. Lizards maintain preferred body temperatures during the day by behavioral means whereas at night body temperatures rapidly fall to ambient. The mean preferred body temperature in *Anolis* is 32°C so the lizards would be expected to maintain this temperature during the day when melatonin levels would be

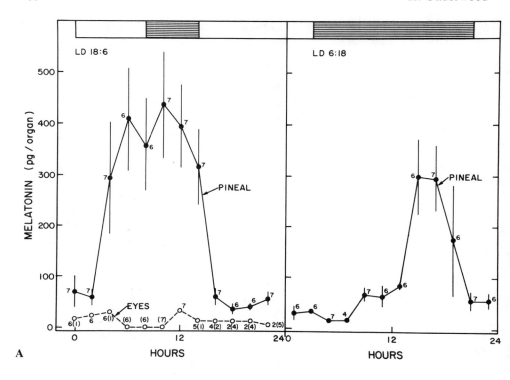

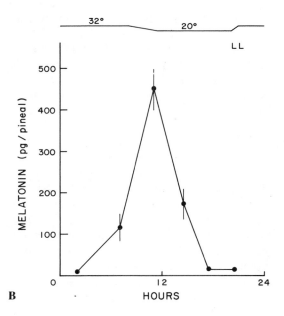

Figure 3-5. Pineal melatonin levels (mean ± SE) of anoles exposed to various light and temperature cycles for several weeks. (A) Anoles entrained to either 18L:6D or 6L:18D (constant 32°C). (B) Anoles entrained to a temperature cycle (12 h at 32°C/12 h at 20°C) and constant light. (C) Anoles exposed to simultaneous light (12L:12D) and temperature cycles (32°C/20°C). In (A) and (C) the numbers beside the mean melatonin levels denote the number of individuals sampled with detectable melatonin levels and, in parentheses, the number with non-detectable levels. In (B) 7 or 8 lizards were sampled at each phase.

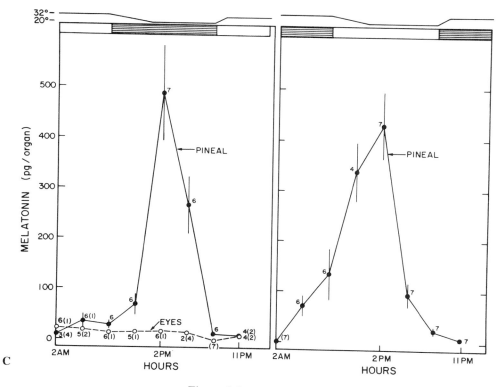

Figure 3-5 *continued*.

low, but at night, when both ambient and body temperatures are low, melatonin levels would be high. In fact, as expected, melatonin levels are high either during the dark phase of an LD cycle or during the cool phase of a temperature cycle (Figure 3-5A,B). Just as differences in the length of the photoperiod can affect several aspects of the pineal melatonin rhythm (Figure 3-5A), the length of the "thermoperiod" can also affect the pineal melatonin rhythm. When entrained to a temperature cycle comprised of 6 h at 20°C and 18 h at 32°C, the amplitude of the PMR in *Anolis* is higher and the time of the peak relative to the 32°C/20°C transition occurs earlier than the melatonin peak of anoles entrained to a temperature cycle of 6 h at 32°C and 18 h at 20°C (Underwood and Calaban 1987a).

Information on the relative strengths of temperature and light as entraining agents can be obtained by "conflicting" the phase of the light and temperature cycles. When exposed to 12L:12D cycles in which the cool phase of a 32°C/20°C temperature cycle occurs during the day, the PMR remains phase-locked to the middle of the cool phase of the temperature and consequently peaks at mid-day (Figure 3-5C) (Underwood 1985a). If the amplitude of the temperature cycle is lowered to 5°C (32°C/27°C) the peak still occurs during the cool phase of the temperature cycle but the peak no longer occurs at mid-day but occurs

3 h earlier. If the amplitude of the temperature cycle is lowered still further, to 2°C (32°C/30°C), the peak occurs at mid-dark of the 12L:12D cycle (Underwood and Calaban 1987a). In summary, under these experimental conditions the temperature cycle was the dominant entraining agent when its amplitude was 12°C, both light and temperature interacted to affect the phase of the PMR when the amplitude of the temperature cycle was 5°C, but the light cycle predominated when the amplitude of the temperature cycle was reduced to 2°C.

The experiments described above show that both light and temperature can act as powerful entraining stimuli and that differences in the length of the daily thermoperiod or photoperiod can affect the phase, amplitude, and duration of the pineal melatonin rhythm. The status of the ambient lighting and temperature conditions, therefore, can be readily translated into an internal signal in the form of the melatonin rhythm. The pineal melatonin rhythm, then, can accurately phase the animal's internal circadian system so that the appropriate event occurs at the "right time of day."

In those higher vertebrates (birds, mammals) examined to date, light suppresses pineal melatonin levels. For example, acute exposure of a bird or a mammal to light at night inactivates the NAT enzyme and causes a rapid fall in melatonin levels (Hamm et al. 1983, Brainard et al. 1983). Depending on species and light intensity and quality, the halving times of melatonin content and NAT activity in response to light at night can vary but are often in the range of 2 to 15 min. In general, the light intensities required to induce suppression of dark-time melatonin levels in diurnal mammals are higher than those required for nocturnal mammals or diurnal birds (Hamm et al. 1983, Brainard et al. 1983). Light, however, does not suppress pineal melatonin levels in *Anolis*. In *Anolis* the PMR can peak during the day (Figure 3-5A) and acute mid-night interruptions by either artificial (24,000 lux) light or sunlight (27,000 lux) does not affect melatonin levels (Underwood 1986b). Clearly, other poikilotherms need to be examined to determine if the insensitivity to the direct suppressive effects of light is confined to some lizards, to all lizards, or to poikilotherms in general.

Since the length of the photoperiod can modify the amplitude and duration of the PMR in *Anolis*, an interesting question is posed: How does the photoperiod modify the shape (i.e., duration) of the PMR if it cannot directly "shape" the rhythm as it does in homeotherms? The answer probably resides in the multioscillator nature of the submammalian pineal organ. The pineal organs of both the lizard *A. carolinensis* and the chicken show circadian rhythms in NAT activity (chicken) and melatonin synthesis (chicken, lizard) (Menaker and Wisner 1983, Takahashi et al. 1980, Hamm et al. 1983). These rhythms are entrainable by light in vitro, and in the chicken it has been shown that a light-entrainable melatonin rhythm can be observed in a dispersed pineal cell culture (Deguchi 1979). Fragments of chick pineal organs containing less than one percent of the whole gland are rhythmic (Takahashi and Menaker 1984). These data suggest that the chicken pineal is comprised of a population of circadian oscillators and also suggest that the photoreceptive input pathway, the circadian clock, and the melatonin output pathway are all located within the same cell. If, as seems likely, the *Anolis* pineal is also comprised of multiple circadian oscillators (individual cells?) which are normally mutually coupled together, the mutual phase relationships among these oscillators will determine the duration and amplitude of the output of these oscillators (i.e., melatonin). Since light is an effective entraining stimulus for circadian oscillators, changes in the duration of the photoperiod would be reflected in altered phase relationships among the coupled oscillators which, in turn, would generate changes in the phase, duration, and amplitude of the melatonin rhythm. According to this hypothesis, light controls pineal melatonin levels indirectly via its effect on the entrainment of circadian oscillators rather than directly via its acute effects on NAT activity.

There is evidence that the duration of the nighttime pineal melatonin rhythm in mammals might also be under the control of circadian pacemakers. In mammals, light perceived by the eyes controls pineal indoleamine synthesis via a route which involves a direct retino-hypothalamic (RH) connection to a neural circadian pacemaker(s) in the suprachiasmatic nuclei (SCN) which, in turn, project to the pineal via the superior cervical ganglia (Rusak and Zucker 1979). Mammals, therefore, differ from other vertebrates since pineal rhythmicity is driven by a clock located elsewhere. In the rat, Illnerová and Vaněček (1982) have good evidence that one circadian oscillator controls the onset of the daily rise in pineal melatonin levels while another controls the fall in nocturnal melatonin levels. The *duration* of the nocturnal melatonin pulse is a function of the phase relationship between these two oscillators (or sets of oscillators), which is a function of the length of the photoperiod. Control of the duration of the melatonin pulse in mammals, therefore, may be similar to that observed in *Anolis* insofar as circadian clocks are involved. In the mammal, however, light perceived by the eyes has an additional effect; that is, light can acutely suppress melatonin levels so that, normally, melatonin levels are prohibited from rising before dusk and/or prohibited from remaining elevated after dawn.

VI. Photic Inputs to the Circadian System: Photoreceptors as Clocks

The clock mechanism obviously has to be coupled to the cyclic photic environment via photoreceptors. There are several kinds of photoreceptors which have been shown to mediate entrainment in lizards; that is, pineal photoreceptors, extraretinal (extrapineal) receptors, and the lateral eyes.

Extraretinal receptors can mediate entrainment since all blinded lizards tested to date (nine species representing four families) are readily entrainable by LD cycles (Figure 3-6B) (Underwood 1973, Underwood and Menaker 1976). Removal of the pineal organ and parietal eye (if present) of *S. olivaceus*, *S. occidentalis*, or *Coleonyx variegatus* does not prevent entrainment of blinded lizards (Underwood 1973). The extraretinal receptors are quite sensitive since lizards can be entrained to light as dim as 1 lux in intensity (Underwood 1973). These receptors are located in the brain although their number and location

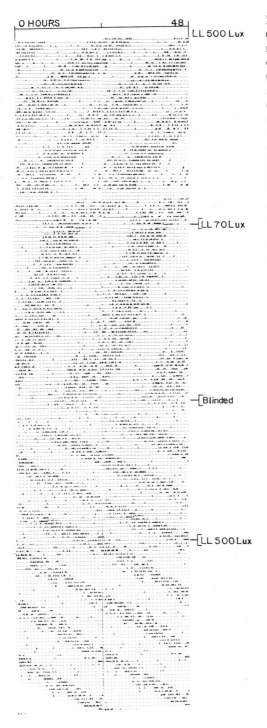

A

Figure 3-6. Effects of blinding (by enucleation) on the activity rhythm of a lizard *Lacerta sicula* exposed to constant illumination (A) or to a light cycle (B). Panel A shows that a sighted lizard obeys Aschoff's rule for a diurnal animal but after blinding the lizard behaves as if it were in constant darkness and cannot discriminate between 70 lux and 500 lux of LL. Panel B gives an example of extraretinal entrainment. The blind lizard entrains to a light cycle comprised of 12 h of 500 lux and 12 h of 70 lux of light per cycle. The upward-pointing arrow indicates the onset of the 500-lux phase of the cycle.

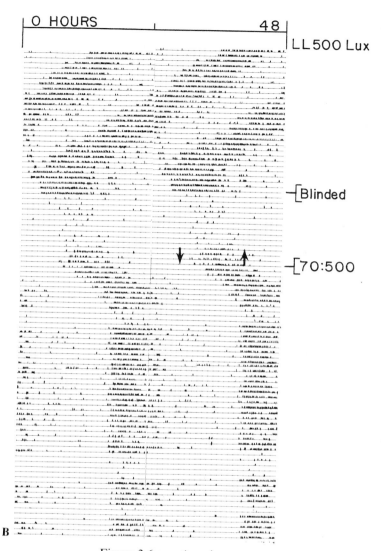

Figure 3-6 *continued.*

are unknown (Underwood 1973, Underwood and Menaker 1976). The hypo-thalamus seems a likely site for at least some of these extraretinal photore-ceptors, however, for several reasons. First, in birds illumination of the medial-ventral hypothalamus can elicit photoperiodic stimulation of the reproductive system. Second, a neural pacemaker (SCN) resides in the hypothalamus, and a close anatomical connection between a clock and its photoreceptors may be advantageous. Third, the diencephalon is known to give rise to the only other identifiable photoreceptors in vertebrates (lateral eyes, pineal system).

Since blinded–pinealectomized lizards are entrainable by LD cycles, the extraretinal receptors must be capable of entraining the "extrapineal" components of the lizard's multioscillator system. Whether or not these extraretinal receptors also have an input into the pineal itself is unknown; however, it is clear that the pineal itself is directly photosensitive in lizards. For example, the lizard pineal contains cells with photosensory appearance, it shows neurophysiological responses to illumination, and the circadian rhythm of melatonin secretion can be entrained by LD cycles in vitro (Hamasaki and Eder 1977, Quay 1979, Menaker and Wisner 1983). In addition, the pineal melatonin rhythm in *Anolis* is entrainable by LD cycles even after removal of the parietal eye and lateral eyes (Underwood and Calaban 1987b). In the intact lizard, light is undoubtedly affecting the circadian system both via extraretinal receptors and via pineal photoreceptors. In addition to these extraocular inputs, the lateral eyes also have an input into the lizard's circadian system. For example, blinding (by enucleation) causes *S. olivaceus* entrained to a very dim green light (0.05-lux intensity) to freerun (Underwood 1973).

The role of the eyes, however, appears to be more complex than that of simply providing additional photic input to the circadian system. To examine the eye's role further, it is necessary to discuss the way in which light can affect a circadian system. Two aspects of light are considered to be potentially important to the entrainment process—the LD or DL *transitions* (termed the nonparametric effect) and the *duration* of the light (termed the parametric effect of light). The light–dark transitions may be the main aspect of the light cycle which elicits entrainment, or the parametric effects of light may be important. For example, parametric entrainment could occur if the light in a LD cycle parametrically increases the velocity of the circadian oscillator and dark decreases its velocity so that an exact 24-h periodicity can be obtained. Continuous lighting conditions have a predictable (parametric) effect on the period of circadian rhythms. In diurnal animals, increases in the intensity of LL shortens the freerunning period and, furthermore, the period of the rhythm in LL is shorter than the period expressed in DD (Aschoff 1979). In nocturnal animals, however, opposite relationships hold between the freerunning period and light. These empirical observations have been termed "Aschoff's rule." In some species (*S. olivaceus, S. occidentalis*) blind lizards can sense the parametric effects of light since the freerunning period changes if the lizard is switched from LL to DD (Underwood and Menaker 1976). In the lizard *Lacerta sicula*, however, the blind lizard cannot discriminate between LL and DD (Figure 3-6A) although it is still entrainable to a light cycle (Figure 3-6B) (Underwood and Menaker 1976). Accordingly, in *Lacerta*, but not in *S. olivaceus* or *S. occidentalis*, extraretinal entrainment must involve a nonparametric mechanism and the eyes must be solely responsible for mediating the parametric effects of light.

Remarkably, blinding *S. occidentalis* by sectioning of the optic nerve actually renders the circadian system more sensitive to the phase-shifting effects of light (Underwood 1985b). Figure 3-2C shows the PRC for blind and sighted *S. occidentalis*; the amplitude of the PRC is severalfold larger in blind lizards. Consistent with this result is the fact that blind lizards reentrain more rapidly to a phase of a LD cycle than sighted lizards and the limits of entrainment are larger in blind lizards (Underwood 1985b). For example, some sighted lizards which freeran when exposed to a 27.5-h (6L:21.5D) or a 29.5-h (6L:23.5D) light cycle entrained following sectioning of the optic nerves. In this species the eyes have an "inhibitory" role in the circadian system.

There are some experimental results in several vertebrate species which indicate that the eyes themselves may be the loci of circadian clocks. A rhythm of melatonin synthesis persists in the eyes of Japanese quail even after the optic nerve has been cut and the superior cervical ganglia removed (Underwood and Siopes 1985, Underwood and Siopes, unpublished). A circadian clock must reside in the eye of the clawed frog *Xenopus* since a circadian rhythm in NAT activity can be observed in vitro (Besharse and Iuvone 1983). If the lizard eye is also the locus of a circadian pacemaker, the concept that the eyes have only a photosensory input into the circadian system will require modification. For example, the increased sensitivity of blind *S. occidentalis* to light might be explained as an effect of removing one oscillatory component from a normally coupled multioscillator system. The remaining components might then show an increased responsiveness to light.

It seems likely that extraretinal photoreception existed before the evolution of the lateral eyes. More primitive eyeless organisms undoubtedly utilized the daily light cycle as an entrainment stimulus. Once the eyes had evolved, they were then "wired" into the system. The fact that extraretinal receptors have persisted shows that they must have some selective advantages or are resistant to alteration. For example, the eyes and extraretinal receptors may be extracting different kinds of information from the photic environment (i.e., parametric versus nonparametric). Or the extraretinal receptors could be an integral part of the clock itself, making them an evolutionarily conserved cell component.

VII. Evolution of Vertebrate Circadian Systems: Pineal and SCN

The mammalian pineal exhibits a host of circadian rhythms in substrate concentrations (i.e., serotonin, melatonin) and enzyme activities (Reiter 1982). These rhythms are under the control of sympathetic nerves, however, and the pineal itself does not exhibit autonomous rhythmicity (Reiter 1982, Rusak 1982). It is widely held that all of the circadian rhythms in mammals, including those in the pineal, are driven by a circadian pacemaker(s) located in the SCN (Rusak and Zucker 1979). Pinealectomy has little effect on circadian rhythms in mammals (Rusak 1982, Underwood 1984).

In birds, the SCN is also important to circadian organization and, although far less studied than in mammals, the avian SCN also seems to act as a circadian pacemaker. For example, lesions of the SCN in birds abolish the circadian rhythm of activity, rendering the birds arrhythmic (Simpson and Follett 1981, Takahashi and Menaker 1982). Pinealectomy can also disrupt circadian rhythmicity in several avian species and, in fact, the effects of pinealectomy in birds (period changes, arrhythmicity) are remarkably similar to those seen in lizards

(Takahashi and Menaker 1979; Underwood 1977, 1981). Organ-cultured chicken pineals can exhibit circadian rhythms in NAT activity or melatonin secretion but these rhythms will persist for only up to four cycles in DD (Takahashi et al. 1980, Kasal et al. 1979). The damping out of these rhythms is probably not due to cellular degeneration since the rhythms will persist for longer periods if the cultures are exposed to 24-h LD cycles (Takahashi et al. 1980). The avian pineal is also innervated by sympathetic nerves from the superior cervical ganglia. One study has shown that removal of these ganglia will disrupt the plasma melatonin rhythm in chickens held in DD (Cassone and Menaker 1983). Taken together, the results indicate that the chicken pineal may only be semi-autonomous; that is, in the absence of input either directly from LD cycles or from the SCN, circadian rhythmicity tends to damp out after a few days in DD. Exogenous melatonin administration can affect circadian rhythmicity (i.e., induce period changes or elicit entrainment) in some avian species (Takahashi and Menaker 1979). Melatonin released from the pineal, therefore, may be involved in controlling extrapineal clocks (in the SCN?). Accordingly, in some birds a feedback relationship might exist from a pineal pacemaker to central clocks (via melatonin) and from the central clocks to the pineal (via nerves).

A role for the SCN in reptiles has not been described. Clearly, however, there do exist extrapineal circadian pacemakers in lizards and one could speculate that some or all of these reside in the SCN. A recent study has demonstrated that the pineal gland of *Anolis* will show a circadian rhythm in melatonin release in vitro which, in contrast to the rhythms exhibited by the chicken pineal, will persist for up to ten cycles in DD (Menaker and Wisner, 1983). There is no information on a role for sympathetic innervation to the lizard pineal. Some sympathetic innervation may occur in lizards but its origin is essentially unknown (Collin and Oksche 1981). In conclusion, there may have been an evolution from the pineal as a relatively autonomous circadian pacemaker in lizards to a system in which the pineal acts as a semi-autonomous pacemaker (birds) or to a system in which the pineal cannot express any rhythmicity on its own (mammals).

References

Aschoff J (1965) Response curves in circadian periodicity. In: Aschoff J (ed) Circadian Clocks. Elsevier/North-Holland, Amsterdam, pp. 95–111.

Aschoff J (1979) Circadian rhythms: influence of internal and external factors on the period measured in constant conditions. Z Tierpsychol 49:225–249.

Besharse JC, Iuvone PM (1983) Circadian clock in *Xenopus* eye controlling retinal serotonin N-acetyltransferase. Nature (London) 305:133–135.

Brainard GC, Richardson BA, King TS, Matthews SA, Reiter RJ (1983) The suppression of pineal melatonin content and N-acetyltransferase activity by different light irradiances in the Syrian hamster: a dose-response relationship. Endocrinology 113:293–296.

Cassone VM, Menaker M (1983) Sympathetic regulation of chicken pineal rhythms. Brain Res 272:311–317.

Collin J-P (1979) Recent advances in pineal cytochemistry. Evidence of the production of indoleamines and proteinaceous substances by rudimentary photoreceptor cells and pinealocytes of amniota. In: Kappers JA, Pévet P (eds) The Pineal Gland of Vertebrates Including Man. Elsevier/North Holland, Amsterdam, pp. 271–296.

Collin J-P, Oksche A (1981) Structural and functional relationships in the nonmammalian pineal organ. In: Reiter RJ (ed) The Pineal Gland, Vol. 1, Anatomy and Biochemistry. CRC Press, Boca Raton, Florida, pp. 27–67.

Deguchi T (1979) A circadian oscillator in cultured cells of chicken pineal gland. Nature (London) 282:94–96.

Hamasaki DI, Eder DJ (1977) Adaptive radiation of the pineal system. In: Crescitelli F (ed) Handbook of Sensory Physiology. Springer, New York, pp. 497–548.

Hamm H, Takahashi JS, Menaker M (1983) Light-induced decrease of serotonin N-acetyltransferase activity and melatonin in the chicken pineal gland and retina. Brain Res 266:287–293.

Illnerová H, Vaněček J (1982) Two-oscillator structure of the pacemaker controlling the circadian rhythm of N-acetyltransferase in the rat pineal gland. J Comp Physiol 145:539–548.

Juillard M-T, Balemans M, Collin J-P, Legerstee WC, Van Benthem J (1983) An in vitro combined radiobiochemical and high-resolution autoradiographic study predicating evidence for indolergic cells in the avian pineal organ. Biol Cell 47:365–378.

Kasal CA, Menaker M, Perez-Polo JR (1979) Circadian clock in culture: N-acetyltransferase activity of chick pineal glands oscillates in vitro. Science 203:656–658.

Lerner AB, Case JD, Takahashi Y, Lee TH, Mori W (1958) Isolation of melatonin, the pineal gland factor that lightens melanocytes. J Am Chem Soc 80:2587.

Lieberman HR (1985) Behavior, sleep and melatonin. In: Wurtman RJ, Waldhauser F (eds) Proceedings of the First International Conference on Melatonin in Humans, 7–9 Nov 1985. Center for Brain Sciences and Metabolism Charitable Trust, Cambridge, ME.

Menaker M, Wisner S (1983) Temperature-compensated circadian clock in the pineal of Anolis. Proc Natl Acad Sci USA 80:6119–6121. ·

Paietta J (1982) Photooxidation and the evolution of circadian rhythmicity. J Theor Biol 97:77–82.

Pittendrigh CS (1965) On the mechanism of entrainment of a circadian rhythm by light cycles. In: Aschoff J (ed) Circadian Clocks. Elsevier/North Holland, Amsterdam, pp. 277–297.

Quay WB (1979) The parietal eye-pineal complex. In: Gans C (ed) Biology of the Reptilia. Academic Press, New York, pp. 245–406.

Ralph CL (1981) Melatonin production by extra-pineal tissues. In: Birau N, Schloot W (eds) Melatonin—Current Status and Perspectives. Pergamon Press, New York, pp. 35–46.

Reiter RJ (ed) (1982) The pineal and its hormones. Progress in Clinical and Biological Research, Vol. 92. Alan R. Liss, Inc., New York.

Rusak B (1982) Circadian organization in mammals and birds: role of the pineal gland. In: Retier RJ (ed) The Pineal Gland: Extra-reproductive Effects. CRC Press, Boca Raton, Florida, pp. 27–51.

Rusak B, Zucker I (1979) Neural regulation of circadian rhythms. Physiol Rev 59:449–526.

Simpson SM, Follett BK (1981) Pineal and hypothalamic pacemakers: their role in regulating circadian rhythmicity in Japanese quail. J Comp Physiol 144:381–389.

Takahashi JS, Menaker M (1979) Physiology of avian circadian pacemakers. Fed Proc 38:2583–2588.

Takahashi JS, Menaker M (1982) Role of the suprachiasmatic nuclei in the circadian system of the house sparrow, *Passer domesticus*. J Neurosci 2:815–828.

Takahashi JS, Menaker M (1984) Multiple redundant circadian oscillators within the isolated avian pineal gland. J Comp Physiol A 154:435–440.

Takahashi JS, Hamm HE, Menaker M (1980) Circadian rhythms of melatonin release from individual superfused chicken pineal glands *in vitro*. Proc Natl Acad Sci USA 77:2319–2322.

Underwood H (1973) Retinal and extraretinal photoreceptors mediate entrainment of the circadian locomotor rhythm in lizards. J Comp Physiol 83:187–222

Underwood H (1977) Circadian organization in lizards: the role of the pineal organ. Science 195:587–589.

Underwood H (1979) Melatonin affects circadian rhythmicity in lizards. J Comp Physiol 130:317–323.

Underwood H (1981) Circadian organization in the lizard *Sceloporus occidentalis*: the effect of pinealectomy, blinding, and melatonin. J Comp Physiol 141:537–547.

Underwood H (1983a) Circadian organization in the lizard *Anolis carolinensis*: a multioscillator system. J Comp Physiol A 152:265–274.

Underwood H (1983b) Circadian pacemakers in lizards: phase response curves and effects of pinealectomy. Am J Physiol 244:R857–R864.

Underwood H (1984) The pineal and circadian rhythms. In: Reiter RJ (ed) The Pineal Gland. Raven Press, New York, pp. 221–251.

Underwood H (1985a) Pineal melatonin rhythms in the lizard *Anolis carolinensis*: effects of light and temperature cycles. J Comp Physiol A 157:57–65.

Underwood H (1985b) Extraretinal photoreception in the lizard *Sceloporus occidentalis*: phase response curve. Am J Physiol 248:R407–R414.

Underwood H (1986a) Circadian rhythms in lizards: phase response curve for melatonin. J Pineal Res 3:187–196.

Underwood H (1986b) Light at night cannot suppress pineal melatonin levels in the lizard *Anolis carolinensis*. Comp Biochem Physiol 84A:661–663.

Underwood H, Calaban M (1987a) Melatonin rhythms in the lizard *Anolis carolinensis* I. Response to light and temperature cycles. J Biological Rhythms (in press).

Underwood H, Calaban M (1987b) Melatonin rhythms in the lizard *Anolis carolinensis* II. Photoreceptive inputs. J Biological Rhythms (in press).

Underwood H, Harless M (1985) Entrainment of the activity rhythm of a lizard to melatonin injections. Physiol Behav 35:267–270.

Underwood H, Menaker M (1976) Extraretinal photoreception in lizards. Photochem Photobiol 23:227–243.

Underwood H, Siopes T (1985) Melatonin rhythms in quail: regulation by photoperiod and circadian pacemakers. J Pineal Res 2:133–143.

Underwood H, Binkley S, Siopes T, Mosher K (1984) Melatonin rhythms in the eyes, pineal bodies, and blood of Japanese quail (*Coturnix coturnix japonica*). Gen Comp Endocrinol 56:70–81.

Chapter 4

Hormones, Behavior, and the Environment: An Evolutionary Perspective

MICHAEL C. MOORE and CATHERINE A. MARLER

I. Introduction

It is widely recognized that successful reproduction requires an organism to synchronize precisely its reproductive physiology and behavior with events in its environment (Farner and Gwinner 1980, Wingfield 1983). This has caused all organisms to evolve interactive links among these factors.

Studies of the links between physiology and behavior in vertebrates have emphasized the endocrine system, particularly the gonadal sex steroid hormones, as a final common pathway for the physiological and environmental factors acting on behavior-regulating centers of the central nervous system. Two lines of evidence support this view. First, circulating levels of sex steroid hormones are sensitive to a variety of factors in both the external and internal environments (Wingfield and Ramenofsky 1985); levels of these hormones are often depressed by stressful conditions or elevated by favorable conditions, presence of suitable mates, etc. (Wingfield and Ramenofsky 1985). Second, the behavior-regulating centers are sensitive to these changes in circulating levels of sex steroid hormones (Arnold and Breedlove 1985, Crews and Moore 1986); in many species, reproductive behaviors are not expressed if gonadal sex steroid hormones are experimentally removed from the blood via castration. Together, these observations have led to a general model in which changes in the external and internal environments are seen to be transduced into behavioral changes by their effects on circulating levels of sex steroid hormones.

Recently, however, several vertebrates in which reproductive behaviors are not dependent on gonadal sex steroid hormones have been documented (Crews 1984, Crews and Moore 1986). These examples are generally found in naturally occurring species with diverse reproductive strategies. Selection for these diverse reproductive strategies has apparently been accompanied by selection for a corresponding diversity of behavior-controlling mechanisms. These observations have led to considerable controversy over the extent to which behavior-controlling mechanisms are evolutionarily "plastic." For example: Is gonadal sex steroid hormone dependence the rule or the exception for vertebrate reproductive behaviors? Is independence from gonadal sex steroid hormone control only a rare occurrence in animals with bizarre breeding strategies or is it more common than previous research has led us to believe?

We believe that, as is the case for many scientific controversies, this one will be resolved not so much by answering these questions as by realizing that these are not the correct questions to ask. It is not correct to dichotomize vertebrate species into those that have gonadal sex steroid hormone-dependent reproductive behaviors and those that do not. Rather, the observation of gonadal sex steroid hormone-independent reproductive behaviors in some species emphasizes a more general point: all species possess a variety of physiological mechanisms that impinge on the behavior-regulating centers. Some of these effects are mediated by changes in circulating levels of sex steroid hormones and others act by alternative pathways (Wallen 1982, Feder 1984, Wingfield and Ramenofsky 1985).

From this perspective, species with gonadal sex steroid hormone-independent reproductive behaviors are only extremes on a continuum. Their behavior-regulating centers simply give less weight to the gonadal sex steroid hormone-mediated pathways than do those of other species. In this context, species with diverse behavior-regulating mechanisms are important to study, not just because they illuminate evolutionary pressures acting on behavior control mechanisms, but because they allow direct investigation of pathways that are present, although frequently obscured, in nearly all species.

In the following sections, we first review the best-documented examples of species that rely extensively on sex steroid hormone-independent mechanisms to activate reproductive behaviors. We then develop the above model in more detail. Finally, we report our recent studies supporting the view that all species possess both gonadal sex steroid hormone-mediated and nonhormonally mediated pathways. By manipulating sex steroid hormone levels in free-living animals, we have demonstrated that seasonal changes in reproductive behavior cannot be accounted for entirely by changes in circulating levels of sex steroid hormones.

II. Examples of Diversity in Behavior-Controlling Mechanisms

A. Red-Sided Garter Snake

The courtship and copulatory behavior of male red-sided garter snakes is the most thoroughly documented example of gonadal sex steroid hormone-independent reproductive behavior (Gartska et al. 1982, Crews et al. 1984, Crews 1984). A large variety of hormone manipulations have been tried with no effect. The failure of the behavior-regulating centers in male garter snakes to use gonadal sex steroid hormones as a signal for activating mating behavior appears to be related to their unusual reproductive strategy. Males produce sperm in late summer and store it over winter in the vas deferens. Mating occurs in the spring shortly after emergence from hibernation. Males have completely regressed and inactive gonads and depend entirely on the sperm stored from the previous summer's gametogenesis.

Thus, the surge in circulating levels of sex steroid hormones that accompanies spermatogenesis occurs at an inappropriate time to be useful as a signal for activating male courtship and copulatory behaviors. Male garter snakes have therefore evolved reliance on alternative sources of information. Current evidence suggests that the behavior-regulating centers of male garter snakes respond directly to the increase in temperature associated with emergence from hibernation to activate courtship and copulatory behavior (Friedman and Crews 1985).

B. White-Crowned Sparrows

As in the red-sided garter snake, copulatory behavior in male white-crowned sparrows is not affected by manipulations of gonadal sex steroid hormones (Moore and Kranz 1983). This species is highly photoperiodic and responds to the vernal increase in daylength to activate the reproductive system (Farner and Gwinner 1980, Wingfield 1983). Prior to their first exposure to long days, male white-crowned sparrows will not copulate with receptive females regardless of their hormonal or photoperiodic treatment. After their first exposure to long days, copulatory behavior appears to be permanently activated (cf. Arnold and Breedlove 1985), even if males are castrated prior to this exposure. Previously photostimulated males mount and copulate with receptive females in all hormonal and photoperiodic treatments, even if castrated or held on short days so that their testes are completely regressed and nonfunctional (Moore and Kranz 1983).

Male white-crowned sparrows evolved gonadal sex steroid hormone-independent activation for reasons similar to, but perhaps more subtle than, those operating in the garter snake (Moore 1984, Crews and Moore 1986). However, as with the garter snake, the specifics of the male reproductive strategy are important determinants of the evolution of the behavior-controlling mechanisms.

Unlike the garter snake, male white-crowned sparrows normally mate with females in the spring when testes are enlarged and spermatogenically active. However, the two behaviors most often associated with gonadal sex steroid hormone control, mating behavior and territorial defense, are expressed at different times of the reproductive cycle. This independent expression would not be possible if both depended on the same hormonal signal.

In white-crowned sparrows, the behavior that is initiated by the male, territorial defense, is tightly linked to gonadal sex steroid hormone control (Moore 1984, Wingfield 1984). The behavior that is initiated by the female, copulatory behavior, has been evolutionarily freed from hormonal control in the male. This emancipation is possible because, unlike territorial behavior, male copulatory behavior only occurs in response to a specific social signal, female solicitation displays. Even though the male is in a state of sexual responsiveness year round, he only expresses copulatory behavior during the brief periods of the breeding cycle when females are sexually receptive. Thus, freeing male copulatory behavior from hormonal control has no evolutionary consequences; this behavior is still expressed only during a precisely defined time of year. The same would not be true of the self-initiated territorial behavior. Expression of this behavior during the nonbreeding season appears to have evolutionary costs (Wingfield 1984, Wingfield and Ramenofsky 1985) that have maintained its link to gonadal sex steroid hormone levels.

C. Parthenogenetic Whiptail Lizards

Several species of lizards in the genus *Cnemidophorus* consist entirely of females that reproduce by true parthenogenesis (see review by Crews and Moore 1987). Despite the fact that females of these species have evolutionarily lost the requirement for males to contribute sperm, they have apparently retained the ability of their ovarian cycles to be stimulated by male copulatory behaviors (Crews and Moore 1987). As a result, females of these species "pseudocopulate" with one another and go through alternate female-like and male-like behavioral phases during the progress of the reproductive cycle.

Female parthenogenetic *Cnemidophorus* typically express male-like behavior during the postovulatory phase of the cycle. Expression of male-like copulatory behavior by a female does not require the induction of a male-like endocrine state. Parthenogenetic females have cycles of ovarian steroid secretion that are virtually identical to those of their sexual ancestral species (Moore et al. 1985, Moore and Crews 1986). Expression of male-like copulatory behaviors in parthenogenetic females is not naturally activated by androgen as it is in males of the ancestral species (Lindzey and Crews 1986), but is instead activated by the high levels of progesterone that exist, as in most oviparous female vertebrates, during the postovulatory phase (Moore et al. 1985, Grassman and Crews 1986). Apparently the preexisting postovulatory surge in progesterone occurs at a point in the cycle when it is adaptive to express male-like behavior and it has been evolutionarily co-opted as a signal.

This illustrates the evolutionary plasticity of behavior-controlling mechanisms because the same behavior is controlled by different hormonal signals in two closely related species. A similar situation appears to pertain in birds that have sex role-reversed mating systems (Fivizzani et al. 1986).

III. Information Flow Between the Environment and the Behavior-Regulating Centers

The above examples and several others indicate that a variety of hormonal, social, and environmental signals activate reproductive behaviors in adult vertebrates (Crews and Moore 1986). These observations support the notions that behavior-regulating mechanisms are evolutionarily plastic and that a variety of pathways, both sex steroid hormone-mediated and nonhormonally mediated, can affect the behavior-regulating centers of the central nervous system. Before we can address the important question of why different mechanisms have evolved in different species, we need a means of conceptualizing differences among species.

All complex behaviors in vertebrates are affected by a variety of factors impinging on the behavior-regulating centers from the environment. These include factors from the abiotic environment (such as temperature and photoperiod), the biotic environment (food availability), the social environment (behavior of mates or intrasexual rivals), and the internal environment (reproductive condition, stress, nutritional plane and availability of other limiting nutrients).

As mentioned in the introduction, it has commonly been assumed that these factors influence reproductive behavior through their effects on circulating levels of sex steroid hormones (Figure 4-1, final common pathway model). Although this is certainly one important pathway by which environmental factors affect the behavior-regulating centers, we believe its importance has been overemphasized. This model is especially weak because it does not easily accommodate the demonstrated evolutionary plasticity. It requires evolution of a completely new, alternative final common pathway to account for diversity among species.

An alternative model that easily accommodates evolutionary plasticity is one in which sex steroid hormones are only one of many pathways for environmental information to affect the behavior regulating centers (Figure 4-1, multiple pathway model). In this model, evolutionary plasticity does not require complete reorganization or substitution of major pathways; it simply requires that the integrating mechanisms of the behavior-regulating centers weight the various inputs differently in different species.

The study of the physiological mechanisms regulating behavior thus becomes not simply one of how environmental factors affect sex steroid hormone levels, but one of how multiple sources of information are integrated by the behavior-regulating centers.

McClintock (1983), Feder (1984), Wingfield and Ramenofsky (1985), and others have pointed out that the misconception about an overly deterministic relationship between hormones and the expression of behavior may have arisen from experiments done under ''controlled'' conditions in the laboratory. In attempts to tease apart the mechanisms by which hormones influenced behavior, all other potential sources of behavior-modifying information were appropriately excluded from or optimized in the experimental design. Results of these studies were then inappropriately generalized to an overly deterministic general model of how hormones affect behavior. More recent studies (see below) done in the presence of multiple sources of information indicate that the social and environmental situation interacts in a complex way with hormonal signals to produce the final expression of a particular behavior. An intriguing possibility is that not only are multiple sources of information capable of influencing the behavior-regulating centers, but that there may be a redundant set of cues.

If the cues are redundant, then manipulations of single cue when others are present may have little effect. It is possible that under the appropriate circum-

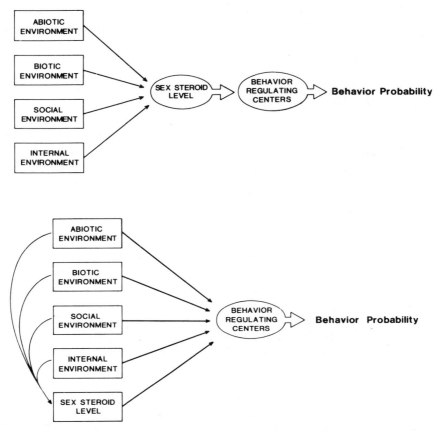

Figure 4-1. Two possible models for conceptualizing the neuroendocrine pathways by which changes in the organism's internal and external environment affect the expression of reproductive behaviors. In the final common pathway model (top), all factors in the internal and external environment affect the behavior-regulating centers of the central nervous system via their eventual effects on circulating levels of sex steroid hormones. This model generates the incorrect prediction that behavioral changes should always closely parallel changes in hormone levels and does not easily accommodate evolutionary plasticity. These observations lead to the second model, the multiple pathway model (bottom). In this model, the sex steroid hormone pathway is only one of several that impinge on the behavior-regulating centers. Although sex steroid hormone levels will have a strong effect on behaviors, the degree of their effect will depend on the situation and the species.

stances, even those species classically thought to have sex steroid hormone-dependent reproductive behaviors may appear to express them independent of sex steroid hormone influence. This possibility is consistent with the observations of Crews (1979), Christie and Barfield (1979), McClintock (1983), Wallen (1982), Ramenofsky (1984), and others that the behavioral effects of hormone manipulations are highly dependent on the situation. For example, whether castration abolishes territorial behavior in both *Anolis* lizards (Crews 1979) and laboratory rats (Christie and Barfield 1979) depends on whether the animals are returned to their home cage. Thus, in one situation (return to novel cage) the behavior appears to be sex steroid hormone dependent, whereas in another situation (return to home cage) the behavior appears to be sex steroid hormone independent. Similarly, in female rhesus macaques, the apparent dependence of sexual receptivity on hormonal state is a function of the social situation (Wallen 1982). In small cages, where the male appears to control the interaction, female sexual behaviors are independent of hormonal state. However, in large enclosures, where the female has a wider repertoire of responses available to her, the frequency of receptive behaviors is dependent on hormonal state.

These examples illustrate that even when the hormonal condition of two animals is identical, the final behavioral response depends on other factors in the environment that influence the behavior-regulating centers independent of effects on sex steroid hormones. It is clear that if we are to understand the role that hormones and other physiological mechanisms play in regulating the expression of behavior, studies must be carried out in the presence of the entire spectrum of environmental information, preferably on free-living animals in their natural environment.

IV. Hormone Manipulations in the Natural Environment: A Case Study

A. General Issues

The above considerations caused us to focus our current research on two critical questions: (1) what is the role of hormones in regulating behavior under natural conditions and (2) what are the evolutionary factors that have led some species to rely more heavily on sex steroid hormone-mediated pathways to regulate their reproductive behavior?

Answering the first question clearly requires experimental manipulations of hormones in free-living animals. There have been numerous recent studies of the effects of artificially increased hormone levels on the behavior of free-living animals (e.g., reviews in Moore 1984, Wingfield and Ramenofsky 1985) but only one limited qualitative study of the effects of castration in a free-living animal (Lincoln et al. 1972). Clearly there is a need for quantitative studies of the effects of both increasing and decreasing hormone levels in the presence of other natural cues. Is it, as predicted by the single final common pathway model, possible to explain all changes in an animal's reproductive behavior as the result of changes in circulating levels of sex steroid hormones or, when other cues are present, will the effects of hormone manipulations be less pronounced or even absent?

Answering the second question posed above about evolutionary factors requires careful selection of study organisms whose life histories have features that will allow separation of the effects of different possible selective factors.

The species studied so far (Crews and Moore 1986, also Section II above) point to three selective factors as possible evolutionary determinants of the degree of sex steroid hormone dependence.

First is the degree to which the timing of the physiologically important changes in sex steroid hormones corresponds with the most appropriate time to express the behavior (garter snake and whiptail examples). Second is the need to coordinate multiple, independently expressed social behaviors that constitute a complex reproductive strategy (white-crowned sparrow example). Third is whether the behavior is a response to an external physical or social stimulus or is self-initiated (white-crowned sparrow example). The latter may be especially important in evolutionarily resolving conflicts arising because of the second situation.

B. The Study Organism and Rationale for the Experiments

With these considerations in mind we chose one of the large iguanid lizards, *Sceloporus jarrovi,* for our current series of experiments. This is a diurnal, territorial species that is very easy to work with in the field. Free-living individuals are oblivious to people, easily caught, and easily observed and have discrete, easily quantified social behaviors. In addition, this species has an unusual reproductive strategy that provides a natural experiment in which the above evolutionary factors make contradictory predictions.

S. jarrovi is a fall breeder; females are pregnant over winter and give birth to live young in early summer; males are territorial during both the summer nonbreeding phase and the fall breeding season, although the intensity of defense is less in the summer than in the fall (Ruby 1978, 1981). During the winter they do not hibernate, but remain active in large aggregations on rock faces with sunny, southern exposures. The annual cycle of males can thus be conveniently divided into three phases: the *winter phase* when males are not territorial and not reproductively active, the *summer phase* when males are territorial and not reproductively active, and the *fall phase* when males are both territorial and reproductively active.

Activation of territorial behavior months before significant gonadal growth occurs presents an interesting test of the influence of the evolutionary forces on behavior-regulating mechanisms. In male *S. jarrovi,* the expression of both sexual and territorial behaviors varies seasonally and both are initiated by the male. This should select for strong sex steroid hormone dependence. However, these two behaviors are expressed at least somewhat independently during the year: territorial behavior is expressed in both summer and fall, whereas sexual behavior is expressed only in the fall. This should select for one of them to be relatively sex steroid hormone independent. Finally, because the activation of territorial behavior precedes the activation of the reproductive system by several months, this should select for relative sex steroid hormone-independent activation of territorial behavior.

To discriminate among these possibilities we performed the following experiments: (1) measurement of circulating levels of sex steroid hormones throughout the year to correlate with changes in expression of social behaviors; (2) castration of free-living males during October at the height of the breeding season when gonadal steroid hormones would be expected to have their maximum influence on behavior; (3) implantation of males with testosterone during the summer to determine if mimicking high breeding season levels of sex steroid hormones would mimic high breeding season levels of behavior; and (4) castration of males during the summer territorial phase to determine the extent to which circulating gonadal steroids influence this nonbreeding season behavior. Behavioral testing in all experiments was done 3 to 4 weeks after hormone manipulations. The effectiveness of the castrations was confirmed by measuring plasma levels of testosterone. Levels were either undetectable or well below those of winter aggregated males.

C. Results

The measurements of sex steroid hormone levels of males throughout the year indicate that seasonal changes in levels of testosterone are tightly correlated with seasonal changes in intensity of territorial behavior (Moore 1986). Testosterone is low in the winter, moderately elevated in the summer, and very high during the fall breeding season. This is consistent with the hypothesis that territorial behavior depends directly on testosterone and is being activated by a reproductive hormone during the nonbreeding season.

These possibilities were tested in a series of castration/hormone replacement experiments under natural conditions. During October, castration had a dramatic effect on several components of territorial and sexual behavior (Moore 1987; Figure 4-2). This confirms that the gonadal sex steroid hormone pathway is important even in the presence of other natural stimuli. Nevertheless, it is equally important that castrated males maintained their territories and gave social displays, although at a much reduced level. Thus, castration reduced the intensity of territorial defense but did not abolish territorial behavior. The castrates did not revert to a nonterritorial winter condition, suggesting that other cues in the environment are capable of maintaining low-level territorial behavior in the absence of sex steroid hormone signals.

The results of mimicking breeding season levels of testosterone by giving silastic implants of testosterone to males in summer (Moore and Marler 1987) produced results that were also consistent with the interpretation that sex steroid hormones are only one of several cues regulating territorial behavior. The implants significantly increased several measures of territorial behavior (Figure 4-2), but the levels attained were still far below those typical of intact males during the height of the breeding season. For example, display frequencies in testosterone-treated males were only 25 to 35% of those measured in males during the fall breeding season. Again, hormone manipulations changed the behaviors in the predicted direction, but are not sufficient by themselves to produce the complete transition to breeding season levels of behavior. Other

Behavior	October Castration	June T−implants	June Castration
Occupancy	0	↑	0
Patrolling	↓	0	0
Head Bobs	0	0	0
Pushups	↓	↑	0
Shudders	↓	↑	0
Fullshows	0	↑	0
Aggressive Response	↓	↑	0
Sexual Response	↓		

Figure 4-2. Schematic summary of the results of the three experiments discussed in the text. A zero indicates no significant difference between experimentals and sham-operated controls, an arrow indicates a significant difference, and the direction of the arrow indicates the direction of the difference (a down arrow meaning experimentals decreased relative to controls, an up arrow meaning an increase relative to controls).

cues, perhaps changes in photoperiod, temperature, or rainfall, apparently interact with changes in hormone levels to produce full expression of breeding season behavior.

Finally, the castration of males during the summer provided results contrary to the hypothesis suggested by the correlations of behavior and hormone levels (Moore and Marler 1987). Castration during the summer, despite the fact that it successfully suppressed testosterone to levels below those of winter males, was completely without effect on several measurements of territorial behavior (Figure 4-2). There is thus no evidence of any involvement of testosterone in maintaining nonbreeding season territoriality. It is still possible that testosterone is involved in the transition from aggregation to territorial behavior early in the spring, although these results make that possibility unlikely.

In summary, castration in October significantly depressed but did not abolish male territorial behavior. Mimicking fall testosterone levels in summer males with implants significantly increased levels of territorial behavior but did not successfully induce full breeding season behavior intensity. Castration during summer, even though it successfully depressed testosterone to 5% of normal summer levels, did not abolish territorial behavior.

From the perspective of the evolutionary questions, these results suggest that temporal association of behaviors with the activation of the reproductive system and the requirement for independent expression of multiple behaviors are most important in determining the degree of sex steroid hormone dependence. Only behaviors expressed while the reproductive system is active appear to be strongly influenced by sex steroid hormones. However, the suggestion that the same behavior, in this case territorial behavior, is controlled by different mechanisms in different seasons was unexpected and may force revision of the evolutionary hypotheses. Nevertheless, this result emphasizes that the behavior-regulating centers will apparently evolve a response to preexisting hormonal signals, but, when hormone signals are not present to coordinate physiological functions, these centers will evolve responses to alternative cues (cf. arguments in Crews 1984).

V. Conclusions

The results of our studies on *S. jarrovi* reconfirm the importance of sex steroid hormones in regulating behavior even when the full spectrum of natural alternative signals is present. Simultaneously, it is completely impossible to account fully for the observed seasonal changes in behavior by invoking only changes in sex steroid hormone levels as the causal agent. These results are therefore consistent with the hypothesis that changes in sex steroid hormone levels are only one of several physiological pathways that influence the behavior-regulating centers of the central nervous system. They also raise the intriguing possibility that the same behavior may be regulated in the same individuals by different mechanisms during different seasons: territorial behavior of male *S. jarrovi* appears to be strongly influenced by sex steroid hormones during the fall but not during the summer.

The conclusion that the effects of sex steroid hormones on behavior can be modified by other factors is not a novel one and has permeated thinking in the field of behavioral endocrinology since its inception. The experiments on *S. jarrovi* extend our understanding of this process because they are to our knowledge the first careful quantification of the behavioral effects of both sex steroid removal and enhancement under natural conditions. The results suggest that the importance of non-steroid-mediated influences on behavior may have been underestimated. Further research with free-living animals in their natural environment is essential if we are to understand fully the role of the various pathways and the ways that they better adapt organisms to their environment.

Studies of naturally occurring species with diverse reproductive strategies have recently focused attention on the importance of sex steroid hormone-independent pathways in influencing reproductive behaviors. It is becoming increasingly apparent that these species do not represent bizarre exceptions to an otherwise universal pattern of sex steroid hormone dependence. Rather they are extremes on a multidimensional continuum of possible relative weightings of the various pathways affecting reproductive behaviors. Further study of species with strong reliance on sex steroid hormone-independent mechanisms will no doubt prove to be a powerful tool for understanding the nature of these various pathways. Similarly, study of these and other naturally occurring species will contribute to what is clearly a major challenge for the future of behavioral physiology: the unraveling of the mechanisms by which information from these various pathways is integrated in the final expression of a particular behavior.

Acknowledgments. We thank David Crews for stimulating discussions that contributed to the development of many of the ideas presented here, David Vleck for pointing out the advantages of *S. jarrovi* for this type of work and for his continuing advice and assistance, Douglas Ruby for helpful discussion and advice, and Christopher Thompson for valuable comments on an earlier draft. The investigations discussed here were supported in part by two awards from the Faculty Grant-in-Aid Program at Arizona State University; an award from Biomedical Research Support Grant S07 RR07112 awarded to Arizona State University by the Biomedical Research Support Program, Division of Research Resources at the National Institutes of Health; and a Presidential Young Investigator Award DCB-8451641 from the National Science Foundation and a Sigma Xi grant-in-aid.

References

Arnold AP, Breedlove SM (1985) Organizational and activational effects of sex steroids on brain and behavior: a reanalysis. Horm Behav 19:469–498.

Christie MH, Barfield RJ (1979) Effects of castration and home cage residency on aggressive behavior in rats. Horm Behav 13:85–91.

Crews D (1979) Neuroendocrinology of lizard reproduction. Biol Reprod 20:51–73.

Crews D (1984) Gamete production, sex hormone secretion and mating behavior uncoupled. Horm Behav 18:22–28.

Crews D, Moore MC (1986) Evolution of mechanisms controlling mating behavior. Science 231:121–125.

Crews D, Moore MC (1987) Reproductive psychobiology of parthenogenetic whiptail lizards. In: Wright J (ed) Biology of the Cnemidophorus. Allen Press, Lawrence, Kansas, in press.

Crews D, Camazine B, Diamond M, Mason R, Tokarz RR, Gartska WR (1984) Hormonal independence of courtship behavior in the male garter snake. Horm Behav 18:29–41.

Farner DS, Gwinner E (1980) Photoperiodicity, circannual and reproductive cycles. In: Epple A, Stetson MH (eds) Avian Endocrinology. Academic Press, New York, pp. 331–366.

Feder HH (1984) Hormones and sexual behavior. Ann Rev Psychol 34:165–200.

Fivizzani AJ, Colwell MA, Oring LW (1986) Plasma steroid hormone levels in free-living wilson's phalaropes *Phalaropus tricolor.* Gen Comp Endocrinol 62:137–144 (in press).

Friedman D, Crews D (1985) Role of the anterior hypothalamus-preoptic area in the regulation of courtship behavior in the male Canadian red-sided garter snake (*Thamnophis sirtalis parietalis*): lesion experiments. Behav Neurosci 99:942.

Gartska WR, Camazine B, Crews D (1982) Interactions of behavior and physiology during the annual reproductive cycle of the red-sided garter snake (*Thamnophis sirtalis parietalis*). Herpetologica 38:104–123.

Grassman M, Crews D (1986) Progesterone induction of pseudocopulatory behavior and stimulus-response complementarity in an all-female lizard species. Horm Behav 20:327–335.

Lincoln GA, Guiness F, Short RV (1972) The way in which testosterone controls the social and sexual behavior of the red stag deer (*Cervus elaphus*). Horm Behav 3:373–396.

Lindzey J, Crews D (1986) Hormonal control of courtship and copulatory behavior in male *Cnemidophorus inornatus*, a direct sexual ancestor of a unisexual, parthenogenetic lizard. Gen Comp Endocrinol 64:411–418.

McClintock MK (1983) The behavioral endocrinology of rodents: a functional analysis. BioScience 33:573–577.

Moore MC (1984) Changes in territorial defense produced by changes in circulating levels of testosterone: a possible hormonal basis for mate-guarding behavior in white-crowned sparrows. Behaviour 88:215–226.

Moore MC (1986) Elevated testosterone levels during nonbreeding season territoriality in a fall-breeding lizard, *Sceloporus jarrovi*. J Comp Physiol 158A:159–163.

Moore MC (1987) Castration affects territorial and sexual behaviour of free-living male lizards, *Sceloporus jarrovi*. Anim Behav 35:1193–1199.

Moore MC, Crews D (1986) Sex steroid hormones in a bisexual whiptail lizard *Cnemidophorus inornatus*, a direct sexual ancestor of a unisexual pathenogen. Gen Comp Endocrinol 63:424–430.

Moore MC, Kranz RH (1983) Evidence for androgen independence of male mounting behavior in white-crowned sparrows. Horm Behav 17:414–423.

Moore MC, Marler CA (1987) Effects of testosterone manipulations on nonbreeding season territorial aggression in free-living male lizards, *Sceloporus jarrovi*. Gen Comp Endocrinol 65:225–232.

Moore MC, Whittier JM, Crews D (1985) Sex steroid hormones during the ovarian cycle of an all-female, parthenogenetic lizard and their correlation with pseudosexual behavior. Gen Comp Endocrinol 60:144–153.

Ramenofsky M (1984) Endogenous plasma hormones and agonistic behavior in male Japanese quail, *Coturnix coturnix*. Anim Behav 32:698–708.

Ruby DE (1978) Seasonal changes in territorial behavior of the iguanid lizard *Sceloporus jarrovi*. Copeia 1978:430–438.

Ruby DE (1981) Phenotypic correlates of male reproductive success in the lizard, *Sceloporus jarrovi*. In: Alexander RD, Tinkle DW (eds) Natural Selection and Social Behavior. Chiron Press, New York, pp. 97–107.

Wallen K (1982) Influence of female hormonal state on rhesus sexual behavior varies with space for social interaction. Science 217:375–377.

Wingfield JC (1983) Environmental and endocrine control of avian reproduction: an ecological approach. In: Mikami S, Homma K, Wada M (eds) Avian Endocrinology: Environmental and Ecological Perspectives. Japan Scientific Societies Press, Tokyo, pp. 265–288.

Wingfield JC (1984) Androgens and mating systems: testosterone-induced polygyny in normally monogamous birds. Auk 101:665–671.

Wingfield JC, Ramenofsky M (1985) Testosterone and aggressive behaviour during the reproductive cycle of male birds. In: Gilles R, Balthazart J (eds) Neurobiology. Springer-Verlag, Berlin, pp. 92–104.

Chapter 5

Circadian Locomotor Rhythms in Birds Correlate with and May Be Explained by Rhythms in Serotonin, *N*-Acetyltransferase, and Melatonin

Sue Binkley

I. Introduction

In the material that follows, I will outline my view of avian pineal function developed in over twenty years of exploration. I will use the results from two species with which I have worked extensively to illustrate the view: sparrows and chicks. Sparrows, *Passer domesticus,* have easily monitored perching behavioral rhythms, but they are less suited for biochemical measurements because they have small pineal glands, possess variable histories, and must be wild-trapped, which limits the numbers available. Chicks, *Gallus domesticus,* are suitable experimental subjects for pineal biochemistry because they are available in large numbers, they are homogenous in age, they have large pineals, their history can be defined in the laboratory, and they are inexpensive; however, locomotor rhythms in chickens are less easily measured in the laboratory and are not as well defined as the perching rhythms of sparrows. The evidence I will develop involves, first, correlative data, second, causal experiments, and third, models to explain the results. Circadian rhythm terminology will be as defined by Aschoff (1965).

II. Correlations

A. Anatomical Correlates

There are correlates in pineal anatomy—pineals of sparrows and chicks possess pinealocytes that appear to be modified photoreceptor-like cells; the pineals of the two species are similarly shaped and located in the top of the head (chicks, Boya and Zamorano 1975; sparrows, Oksche and Kirchstein 1969).

B. Biochemical Correlates

Both sparrow and chick pineals are capable of melatonin synthesis which occurs in the dark-time of a light–dark cycle. The biochemical cycles are negatively correlated with locomotor activity which occurs in the light-time of a light–dark cycle in both chicks and sparrows. The regulation of the rhythms in pineal biochemistry and in locomotor activity is similar.

1. Melatonin Synthesis

Part of the pathway for biosynthesis and metabolism of melatonin includes melatonin synthesis from serotonin by the enzymes NAT (*N*-acetyltransferase) and HIOMT (hydroxyindole-*O*-methyltransferase). NAT and melatonin exhibit dark-time peaks, while HIOMT has but small daily changes (Figure 5-1). The conclusion drawn was that the melatonin rhythm is a consequence of the rhythm in NAT (Binkley et al. 1973). Throughout this paper, NAT data will be used for illustration; in fact, the rhythmic changes seen would also be found in melatonin (pineal, cerebrospinal fluid, blood, urine).

2. Entrainment by Light–Dark Cycles

Locomotor activity occurs in the daytime; melatonin is high at night. Both rhythms *entrain* to (synchronize with) environmental light–dark cycles. As an illustration, both rhythms reverse *phase* when the lighting schedule is *reversed* by extending the light or dark by 12 h (a 180° phase shift) (Figure 5-2). The response to reversal correlates negatively for the sparrow locomotor activity and chick NAT.

The shape of both locomotor and NAT rhythms is a function of *photoperiod*. In 12L:12D, *bimodality* (the presence of a first peak just after dawn and a second peak just before dusk; Binkley 1976a) occurs in pineal serotonin and in perch-hopping activity (Figure 5-3). The number of birds that show bimodal patterns is maximal (75%) in 16L:8D and decreases as the photoperiod shortens or lengthens (Binkley 1985).

Sparrows and chicks change the phase, duration, and amplitude of their rhythms in response to *photoperiod* (Figure 5-4). When sparrows were subjected to successive lighting treatments where the amount of light per day was varied, they entrained to all LD cycles (from 1L:23D to 23L:1D). In cycles with short light (e.g., 4L:20D), the birds' activity anticipates and trails lights on (Binkley 1978, Binkley and Mosher 1985a). Chick NAT was measured in LD cycles (from 22L:2D to 2L:22D). The daily profile in chick pineal NAT was also modified by photoperiod. NAT rose when lights-out occurred (it was "phase locked" to lights-out). NAT declined before or after lights-off but its *duration* was modified by the length of the dark; the *amplitude* of the NAT cycle was greater in long than in short photoperiods (Binkley et al. 1977). Thus, perch-hopping and NAT exhibit negatively correlated changes in duration.

Figure 5-1. Biochemical measurements in sparrow (top) and chick (bottom) pineal glands harvested serially over 24 h in 12L:12D show similar indole metabolism (Binkley et al. 1973, Binkley 1983b). There were rhythms in serotonin (bimodal), *N*-acetyltransferase (NAT), and melatonin. Each point of the chick data (bottom) represents the average of multiple measurements (two pineals for serotonin, four pineals for NAT and melatonin, six pineals for HIOMT). Each point of the sparrow data (top) represents an average for pineals of four birds; the vertical lines are one standard error. The time scale represents time from the sparrows' point of view (in this case, 1 h of circadian time (CT) = 1 h of Eastern Daylight Savings Time; CT0 is set by convention as the time the sparrows begin locomotor activity, which is coincident with lights-on). Detailed 24 h profiles of sparrow pineal melatonin and HIOMT were not measured in vivo, but these parameters have been measured on pools of sparrow pineal glands: HIOMT was 810 pmoles/pineal/h in the light-time and 756 pmoles/pineal/h in the dark-time (Binkley 1976b); melatonin was 0.33 ± 0.03 ng in the light and 2.11 ± 0.39 ng in the dark (Binkley, unpublished).

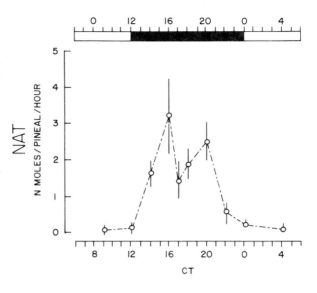

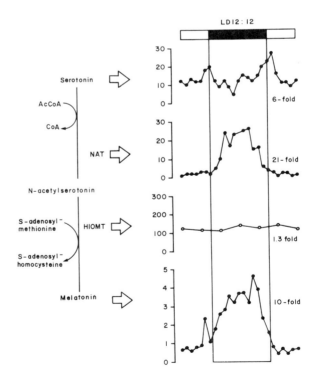

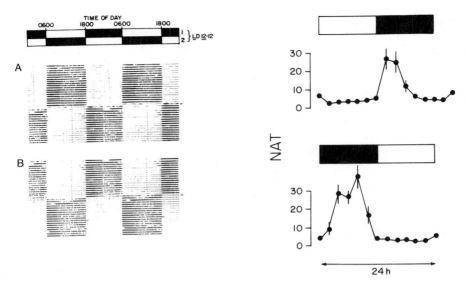

Figure 5-2. Sparrow perching and chick pineal NAT rhythms reentrain when the 12L:12D cycle is reversed (Binkley 1978). (Left) Perch-hopping event records as in Figure 5-4 (except that the records were duplicated horizontally so that the full course of the rhythms can be seen) are shown for two sparrows subjected to reversal of their 12L:12D lighting regimen (represented by bars 1 and 2). (A) is the record of a sparrow that exhibited 7 transients when it was shifted from cycle 1 to 2 by extending the dark. (B) is the record of a sparrow that exhibited 1 transient when it was shifted from cycle 2 to 1 by extending the light. (Right) Chick pineal NAT was measured in the light before (upper) and after (lower) the 12L:12D lighting regimen was reversed (Binkley and Mosher, unpublished).

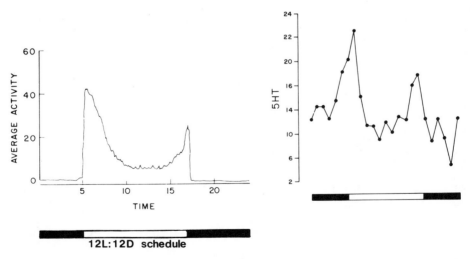

Figure 5-3. A bimodal pattern (left, two peaks/ 24 h) was measured in 12L:12D by counting and averaging sparrow perch-hops over 24 h (Binkley 1976a); a bimodal pattern (right) was also observed in chick pineal serotonin (Brammer and Binkley 1979). Peaks for both measures occur at dawn and dusk.

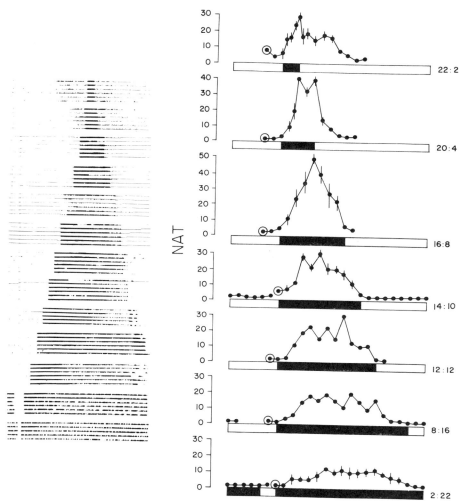

Figure 5-4. Photoperiod alters the pattern of sparrow perch-hopping and chick pineal NAT. (A) Five days of perch-hopping event records are shown from individual sparrows kept in 24-h lighting regimens in a series of photoperiods (top to bottom: 1, 2, 4, 6, 8, 10, 12, 14, 16, 18, 20, 22, 23 h of light per day) (Binkley 1978, 1985, Binkley and Mosher 1985c); each line represents 24 h of recording; the lines are arranged vertically in chronological order. Most of the activity occurred in the light-time, and its duration was a function of photoperiod. (B) NAT was measured in groups of chicks raised in a series of photoperiods (top to bottom: 22, 20, 16, 14, 12, 8, 2 h of light per day; lighting for a given NAT graph is illustrated by the bar below it); the duration of NAT was a function of the length of the dark-time (scotoperiod).

2. Constant Conditions

Sparrow locomotor activity (Figure 5-5) entrains to light–dark cycles (12L:12D), freeruns (persists) in constant dark (DD) with a *period* length that is usually greater than 24 h, freeruns in dim constant light (LL) with a period that is usually less than 24 h, and is aperiodic in bright constant light (Binkley 1977, 1978). The responses do not require the eyes—blind sparrows exhibit the changes; the sparrows have extraretinal light perception (Menaker 1968). Sim-

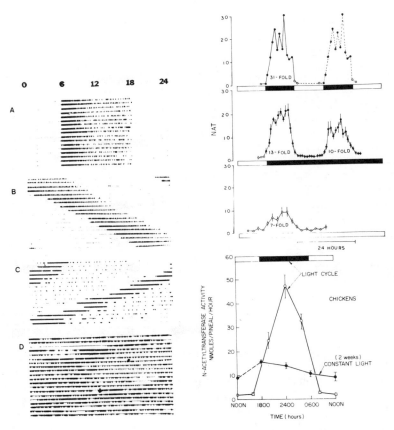

Figure 5-5. Sparrow perching and chick pineal NAT responded to constant dark and light. Four individual records of sparrow perching (each line represents 24 h of event recording) are shown for birds entrained by 12L:12D (A), freerunning in constant dark (B), freerunning in constant dim light (C), and aperiodic in constant bright light (D). Chick pineal NAT profiles are shown for similar conditions (top to bottom): entrained by 12L:12D, freerunning in DD, freerunning with reduced amplitude in dim LL, and after 2 weeks of bright LL.

ilarly, chick NAT is rhythmic in 12L:12D, freeruns in DD (with 32% of LD amplitude), freeruns in dim LL (with 23% of LD amplitude), and bright LL for two weeks abolishes the rhythm in blind or sighted chickens (Ralph et al. 1975). Thus the perch-hopping and NAT have negatively correlated changes in response to constant lighting conditions.

3. Pulses

Circadian rhythms are affected by abrupt changes in lighting conditions—pulses (e.g., 4 h of light or dark) phase shift the timing of the rhythms subsequent to the perturbation by the pulse. Sparrow perch-hopping and chick NAT phase shift responses are similar (Figure 5-6).

Sparrows pretreated with (entrained to) 12L:12D begin activity in DD at a time which extrapolates to their onset in 12L:12D. If a light pulse is imposed during the DD, it may delay or advance the phase of the rhythm. A sparrow placed in dim LL advances its onset; dark pulses imposed in dim LL may advance or delay the phase of the rhythm. There are some times relative to the birds' circadian cycle where light or dark pulses produce no effect; the birds are "insensitive"—there is a "dead zone" (Klein et al. 1985). Chick NAT rhythms also phase shift in response to pulses (Binkley et al. 1981, 1985). Plots of phase-shift responses (to 4-h light or 4-h dark pulses in sparrow perching or chick pineal NAT) versus time of the pulse (phase response curves, PRCs) are similar (Klein et al. 1985, Binkley et al. 1981, Binkley 1983a, 1985, Binkley and Mosher 1985b) (Figure 5-5). The responses to light and dark are roughly opposite to each other for both species.

C. Photoreceptors

Pineal light sensitivity was suggested for both sparrows and chicks. Shielding the pineal region from light affects the reproductive responses of sparrows (Menaker et al. 1970); shielding the pineal region of chicks reduces the pineal response of NAT to light (Binkley et al. 1980a). But shielding techniques are too imprecise for us to conclude that the pineal is light sensitive. Direct evidence for avian pineal light sensitivity is provided by the fact that chick pineal NAT is attenuated by light in vitro (Binkley et al. 1978a, Deguchi 1979, Wainwright and Wainwright 1979).

Extraretinal light perception is present in both sparrows and chicks. The locomotor activity of blinded sparrows is affected by light—they entrain to light–dark cycles and the period length of their freeruns is altered by the intensity of constant light (Menaker 1968, McMillan et al. 1975a,b,c). The locomotor activity of blind chickens is also affected by light—blinded chickens entrain to 12L:12D and they are able to reentrain to a 6-h phase shift (90°) of the cycle (Nyce and Binkley 1977).

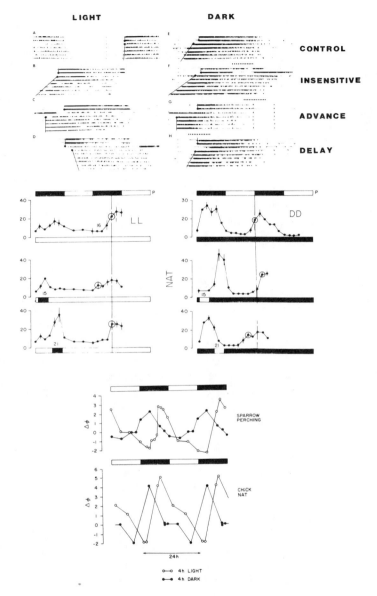

Figure 5-6. Sparrow perching and chick pineal NAT were similarly phase-shifted by 4-h light and dark pulses. Eight examples (top) of individual sparrow perching responses are shown for controls (no pulses) and sparrows given pulses (bars) at different times relative to their previous 12L:12D cycles. Pineal NAT profiles during and following pulses are shown for 6 groups of chickens (middle). Plots of sparrow perch-hopping (upper) and chick pineal (lower) phase shifts versus time of pulse relative to prior 12L:12D (phase response curves duplicated horizontally for ease in comparison; horizontal bar represents 12L:12D for 48 h; data obtained in the type of experiments illustrated in the preceding figures) show similar phase-shifting responses for the two rhythms (Klein et al. 1985, Binkley et al. 1981, Binkley and Mosher 1985b).

III. Causes

A. Melatonin

Sparrows have a rhythm in NAT and they produce melatonin (Binkley 1984) (Figure 5-1). Exogenous melatonin alters rhythms in sparrow behavior: (i) after an injection they assume a roosting posture and drop their body temperature as they do at night (Binkley 1974), (ii) implanted constant-release melatonin capsules abolish the locomotor rhythm or shorten the period of the rhythm (Turek et al. 1976); and (iii) melatonin in the water attenuates the perching rhythm (Binkley and Mosher 1985a). Melatonin in constant-release capsules, however, did not alter the LD cycle in chick pineal NAT or HIOMT, which implies that the pineal is not the site at which exogenous melatonin acts to produce effects (Binkley et al. 1978b).

B. Pinealectomy and Blinding

Pinealectomy abolishes circadian locomotor and temperature rhythms in sparrows kept in DD; blind sparrows exhibit freerunning locomotor rhythms in DD (Gaston and Menaker 1968, Binkley et al. 1972). Pinealectomy did not abolish locomotor rhythms in chickens (MacBride 1973) but blinding abolished DD circadian activity in half of the chickens tested (Nyce and Binkley 1977).

Sparrow eyes have some NAT activity but most is in the pineal (Binkley et al. 1979). Chick eyes and pineal glands have near equal rhythms in pineal NAT and melatonin (Binkley et al. 1979, 1980b). The presence of substantive melatonin synthesis in the eyes may be the reason that pinealectomy fails to cause arrhythmia in chickens. We note that the possibility that eyes and pineal jointly may control locomotor rhythms has been shown in pigeons and quail (Underwood and Siopes 1984, Ebihara et al. 1984).

C. Neural Regulation

Rat pineal glands are clearly regulated by secretion of norepinephrine (NE) in response to signals relayed from the eyes via the hypothalamus. Isoproterenol, which is similar to NE, stimulates rat pineal NAT; superior cervical ganglionectomy abolishes the NAT cycle (reviewed in Binkley 1983b).

This is not the case in birds. Sparrow or chick pineal NAT is not stimulated by norepinephrine in vitro or by injection of isoproterenol in vivo (Binkley 1976b). Moreover, attempts to abolish rhythmicity with chemical sympathectomy were negative in sparrows (Zimmerman and Menaker 1975, 1979). This would seem to suggest the absence of the neural regulatory pathway in birds. However, surgical sympathectomy did not abolish the NAT or melatonin rhythm in pineals of chicks kept in LD but did attenuate the cycle in DD (Ralph et al. 1975, Cassone and Menaker 1983). Additional evidence for some neural regulation, albeit more complex than in the rat, comes from the fact that the eyes can influence chick pineal NAT: the rapid plummet in response to light is slower in blinded chicks and patching the eyes reduces the inhibition of pineal NAT by light extension (Binkley et al. 1975, 1980a).

D. NAT and Melatonin Synthesis In Vitro

Chick and sparrow pineals are capable of rhythmic cycles of NAT and/or melatonin when isolated in static or superfusion organ culture in LD (Binkley et al. 1978a, Wainwright and Wainwright 1979, Kasals et al. 1979, Deguchi 1979, Takahashi 1981, Takahashi et al. 1980).

Evidence of clocklike abilities in the pineal also comes from the fact that chick pineals exhibit "timekeeping" in static organ culture—the performance in culture is dependent upon the time the glands are removed from the chicks (Binkley et al. 1977, Wainwright and Wainwright 1981a, Deguchi 1979). The evidence that the pineals "know what time it is," that they possess phase information, is particularly satisfying because pinealectomized sparrows receiving pineal transplants resume rhythmicity and assume the phase of the donor animal (Zimmerman and Menaker 1979).

E. Pharmacology

The story of avian pineal pharmacology is as yet piecemeal.

Sparrows in DD respond to melatonin (arrhythmia, shortened periods), to the serotonin agonist fluoxetine (phase shifts), and to the serotonin neurotoxin 5,6-dihydroxytryptamine (phase shifts) (Cassone and Menaker 1985).

The response of *chick* pineal NAT to pharmacological agents can be examined by adding drugs to the cultures or by injecting them. Many drugs injected into chicks at the time of lights-out inhibited chick pineal NAT—serotonin, L-dopa, pargyline, histamine, epinephrine, isoproterenol, cycloheximide, and ethanol (Binkley et al. 1978b, 1980a). Norepinephrine and pargyline inhibited chick pineal NAT when added to culture (Binkley 1976b, Riebman and Binkley 1979). Cyclic-AMP-stimulating agents (e.g., dibutryl cAMP, theophylline) stimulate chick pineal NAT when added to cultures (Riebman and Binkley 1979, Wainwright and Wainwright 1981b). Cyclic GMP exhibits cycles that are positively correlated with NAT in chick pineals in vitro (Wainwright 1980).

The common elements in the pharmacology of chick pineal NAT and sparrow perch-hopping rhythms appear so far to be (i) the lack of stimulation by adrenergic agents and (ii) the importance of serotonin and melatonin.

IV. Models and Explanations

We can explain many pineal functions in terms of pineal biochemical rhythms and their control by light and dark.

A. Anatomical Interrelationships

The effects of ablations in birds have yielded a set of "components"—photoreceptors (eyes, extraretinal light receptor, pineal), oscillators (pineal, suprachiasmatic nuclei of the hypothalamus), and neural connections (superior cervical ganglia). These components can be arrayed in hierarchies and feedback systems in a variety of ways but the interrelationships are not as yet entirely understood (Binkley 1980, Cassone and Menaker 1984).

B. Photoperiod Measurement

The circadian, light-sensitive, and refractory properties of NAT suit it (and its output melatonin) for photoperiod measurement. I proposed a model (Binkley 1976c) which divided the control of NAT into two parts that derive from a circadian oscillation:

   (i) a refractory period begins about the middle of the subjective night in 12L:12D and runs to mid-subjective day;
   (ii) a sensitive period forms the other half of the cycle.

If dusk occurs during the sensitive period, NAT rises and continues in dark for a duration that is set by the 12L:12D cycle; if light interrupts the dark during the sensitive period, NAT plummets rapidly and nightlength is transduced to a melatonin signal. Measurement of photoperiod requires comparison with some standard. The system can account for this because the duration standard (which I have called the "timestick") for the NAT refractory and sensitive periods is fixed by the prior 12L:12D cycle.

The change from "sensitive" to "refractory" can explain light-break experiments in photoperiodic control of reproduction. The circadian clock provides a "sensitive" period (a gate, a window) when dark can initiate NAT. This sensitive time encompasses the time of expected lights-out and the early subjective dark-time. Once NAT has initiated, light causes a rapid plummet, and if the subjective night has just begun, NAT can reinitiate, record a "long" night, and suppress the gonads. A light pulse at the very end of the subjective night has little effect; NAT has run its course, a "long" night has been recorded, and the gonads are suppressed. Light pulses in the mid to late subjective night cause a rapid plummet in NAT—NAT cannot reinitiate because the system has become refractory, a "short" night is mimicked (Binkley 1976c, 1983a), and the gonads grow as they would on a long photoperiod. As I interpreted (Binkley 1978) light-break experiments with sparrow reproduction (Menaker 1965), sparrow gonad growth was a function of the length of the dark-time—mid-to-late subjective night light breaks resulted in gonad growth (a short night was measured as would be expected from the NAT response to night interruption by light pulses). Pinealectomy and melatonin have not yet been shown to affect sparrow reproduction, and such demonstrations may require very special conditions of photoperiod, time of administration, and consideration of nonpineal sources of melatonin.

C. Phase Shifting

The circadian, sensitive, and refractory properties of NAT suit it (and its output melatonin) for explaining phase shifting (Figure 5-6). I here draw parallels between the responses of NAT and the phase-shifting responses of circadian rhythms. I suggest that the change from sensitive to refractory is responsible for the change in direction of phase shift of circadian rhythms.

First, a light pulse in the early subjective light-time, in the sensitive period, causes a rapid plummet in NAT but NAT can reinitiate and complete its nightly time program. The phase of the subsequent NAT cycle is *delayed* a small and limited time (related to low values produced during the pulse). Likewise, in PRCs, light pulses in the early subjective night produce delays.

Second, a light pulse in the late subjective night, in the refractory period, causes a rapid plummet in NAT, NAT cannot reinitiate, the clock is reset, NAT begins the course of a subjective day, and the subsequent cycle is thus *advanced*. Likewise, in PRCs, light pulses in the late subjective night produce advances.

Third, a dark pulse in the early subjective night, in the sensitive period, permits NAT to rise. Restoration of light begins a subjective day and the subsequent cycle is *advanced*. Likewise, in PRCs, dark pulses in the early subjective night produce advances.

Fourth, a dark pulse in the late subjective night, in the refractory period, may permit a rise in NAT (depending on how late the pulse was given), and the subsequent cycle is *delayed* because the resumption of light after the pulse begins a new subjective day. Likewise, in PRCs, dark pulses in the late subjective night produce delays.

Fifth, there is a point in the mid-subjective night when the NAT response to dark becomes refractory. Likewise, there is a point in the mid-subjective night when PRCs change from advances to delays (or delays to advances)—I have called this point "xing."

Sixth, there is a "dead zone" in the subjective day because of the refractory period (when NAT cannot be provoked by dark). Likewise, PRCs contain a "dead zone" which falls in the early subjective day.

D. Enzyme Clock

I proposed a model based on the pharmacology, effects of surgery, and effects of light (using data from both rat and chick pineals) (Binkley 1983b), which begins to address the details of how the NAT cycle is generated and the means by which its time program is set. It is described as a sequence of events, as follows.

During the refractory period, begun by lights-on, there is an accumulation of something (I called it an "initiator" and visualize it as a chemical substance or a change in a macromolecule) which permits the initiation of NAT. For example, this might be the synthesis of the RNA for the production of the next night's NAT. This process fixes, or "programs," the duration, amount, and forms (there may be a more stable and a less stable form of NAT) of the NAT of the next cycle.

During the sensitive period, if dark occurs, NAT is synthesized according to the program established in the prior light-time (Binkley and Mosher 1984). The course of NAT thus depends on this program and whether light occurs again. New synthesis of NAT (at least one labile form) must continue throughout the night (because cycloheximide causes NAT to drop whenever in the night it is injected; Binkley et al. 1978b) to keep NAT high; there is a protease that constantly degrades NAT, so that if synthesis is stopped, a rapid plummet occurs.

E. Indole Model

A simple model to consider is an "indole model for control of circadian rhythms in birds":

(i) Serotonin is associated with the provocation of activity and melatonin is associated with the mechanism that elicits rest.

(ii) Light inhibits perching activity in two ways: (a) light directly inhibits cells in the retina and pineal that have NAT and produce melatonin and (b) light perceived via retinal photoreceptors inhibits pineal NAT via a neural route through the suprachiasmatic nuclei and pineal.

(iii) Circadian rhythm time signals are generated in cells of the pineal and suprachiasmatic nuclei (oscillator cells)—this information is distributed by the sympathetic nervous system and by melatonin secretion.

(iv) The shapes of the rhythms of the oscillator cells can be modified by prior light–dark signals; the length of the "timestick" is a function of the length of subjective night that is programmed in the oscillator cells.

(v) The pineal is inhibited by adrenergic agents (e.g., norepinephrine) from the circulation or sympathetic nerve endings.

While much of the evidence I have discussed supports this model, there are some disturbing pieces of information. First, there is the failure of sympathetic interventions to disrupt sparrow locomotor rhythms. Second, there is the puzzling redundancy in photoreceptor and oscillator cell sites. Third, it is likely that a second enzyme, HIOMT, provides seasonal modulation of pineal melatonin output in addition to that resulting from NAT (sparrows, Barfuss and Ellis 1971; chicks, Lauber et al. 1968). Fourth, a target by which melatonin and serotonin may affect rest-activity has not been demonstrated.

Acknowledgments. The research was supported by NSF PCM 8314331 to S.B. and NSF DCB-8613594.

References

Aschoff J (1965) Circadian Clocks. North-Holland, Amsterdam, pp.vii–xix.

Barfuss D, Ellis L (1971) Seasonal cycles in melatonin synthesis by the pineal gland as related to testicular function in the house sparrow, *Passer domesticus*. Gen Comp Endocrinol 17:183–193.

Binkley S (1974) Pineal and melatonin: circadian rhythms and body temperature. In: Scheving L, Halberg F, Pauly J (eds) Chronobiology. Igaku Shoin, Tokyo, pp. 582–585.

Binkley S (1976a) Computer methods of analysis of biorhythm data. In: DeCoursey J (ed) Biorhythms in the Marine Environment. Columbia, South Carolina, University of South Carolina Press, pp. 53–62.

Binkley S (1976b) Comparative pineal biochemistry in birds and mammals. Am Zool 16:57–65.

Binkley S (1976c) Pineal gland biorhythms: N-acetyltransferase in chickens and rats. Fed Proc 35:2347–2352.

Binkley S (1977) Constant light: effects on the circadian locomotor rhythm in the house sparrow. Phys Zool 50:170–181.

Binkley S (1978) Light-to-dark transitions and dark-time sensitivity: importance for the biological clock in the house sparrow. Phys Zool 51:272–278.

Binkley S (1980) Functions of the pineal gland. In: Epple A, Stetson M (eds) Avian Endocrinology. Academic Press, New York, pp. 53–74.

Binkley S (1983a) Circadian rhythm in pineal N-acetyltransferase activity: phase shifting by light pulses II. J Neurochem 41:273–276.

Binkley S (1983b) Rhythms in ocular and pineal N-acetyltransferase: a portrait of an enzyme clock. Comp Biochem Physiol 75A:123–129.

Binkley S (1985) Light and dark control circadian phase in sparrows. In: Brown G, Wainwright S (eds) The Pineal Gland, Endocrine Aspects. Pergamon Press, New York, pp. 59–65.

Binkley S, Mosher K (1984) Prior light alters the circadian clock in the chick pineal gland. J Exp Zool 232:551–556.

Binkley S, Mosher K (1985a) Circadian rhythm in pineal N-acetyltransferase activity: phase shifting by dark pulses III. J Neurochem 451:875–878.

Binkley S, Mosher K (1985b) Oral melatonin produces arrhythmia in sparrows. Experientia 41:1615–1617.

Binkley S, Mosher K (1985c) Direct and circadian control of sparrow behavior by light and dark. Physiol Behav 35:785–797.

Binkley S, Kluth E, Menaker M (1972) Pineal and locomotor activity: levels and arrhythmia in sparrows. J Comp Physiol 77:163–169.

Binkley S, MacBride S, Klein D, Ralph C (1973) Pineal enzymes: regulation of avian melatonin synthesis. Science 181:273–275.

Binkley S, MacBride S, Klein D, Ralph C (1975) Regulation of pineal rhythms in chickens: refractory period and nonvisual light perception. Endocrinology 96:848–853.

Binkley S, Riebman J, Reilly K (1977) Timekeeping by the pineal gland. Science 197:1181–1183.

Binkley S, Riebman J, Reilly K (1978a) The pineal gland: a biological clock in vitro. Science 202:1198–1201.

Binkley S, Riebman J, Reilly K (1978b) Regulation of pineal rhythms in chickens: inhibition of dark-time rise in N-acetyltransferase activity. Comp Biochem Physiol 59C:165–171.

Binkley S, Hryshchyshyn M, Reilly K (1979) N-Acetyltransferase activity responds to environmental lighting in the eye as well as in the pineal gland. Nature (London) 281:479–481.

Binkley S, Reilly K, Hernandez T (1980a) N-Acetyltransferase in the retina II: Interactions of the eyes and pineal gland in response to light. J Comp Physiol 140:181–183.

Binkley S, Reilly K, Hryshchyshyn M (1980b) N-Acetyltransferase in the chick retina I: Circadian rhythms controlled by environmental lighting are similar to those in the pineal gland. J Comp Physiol 139:103–108.

Binkley S, Muller G, Hernandez T (1981) Circadian rhythm in pineal N-acetyltransferase activity: phase shifting by light pulses I. J Neurochem 37:798–800.

Boya J, Zamorano L (1975) Ultrastructural study of the pineal gland of the chicken (*Gallus gallus*). Acta Anat 92:202–226.

Brammer M, Binkley S (1979) Daily rhythms of serotonin and N-acetyltransferase in chicks. Comp Biochem Physiol 63C:291–296.

Cassone V, Menaker M (1983) Sympathetic regulation of chicken pineal rhythms. Brain Res 272:311–317.

Cassone V, Menaker M (1984) Is the avian circadian system a neuroendocrine loop? J Exp Zool 232:539–549.

Cassone V, Menaker M (1985) Circadian rhythms of house sparrows are phase-shifted by pharmacological manipulation of brain serotonin. J Comp Physiol A 156:145–152.

Deguchi T (1979) Circadian rhythm of serotonin N-acetyltransferase activity in organ culture of chicken pineal gland. Science 203:1245–1247.

Ebihara S, Uchiyama K, Oshima I (1984) Circadian organization in the pigeon, *Columba livia*: the role of the pineal organ and the eye. J Comp Physiol A 154:59–69.

Gaston S, Menaker M (1968) Pineal function: the biological clock in the sparrow? Science 160:1125–1127.

Kasals C, Menaker M, Perez-Polo J (1979) Circadian clock in culture: N-acetyltransferase activity of chick pineal glands oscillates in vitro. Science 203:656–658.

Klein S, Binkley S, Mosher K (1985) Circadian phase of sparrows: control by light and dark. Photochem Photobiol 41:453–457.

Lauber J, Boyd J, Axelrod J (1968) Enzymatic synthesis of melatonin in avian pineal body: extraretinal response to light. Science 161:489–490.

MacBride S (1973) Pineal biochemical rhythms of the chicken: light cycle and locomotor activity correlates. Ph.D. thesis, University of Pittsburgh.

McMillan J, Keatts H, Menaker M (1975a) On the role of eyes and brain photoreceptors in the sparrow: entrainment to light cycles. J Comp Physiol 102:251–256.

McMillan J, Elliott J, Menaker M (1975b) On the role of eyes and brain photoreceptors in the sparrow: Aschoff's rule. J Comp Physiol 102:257–262.

McMillan J, Elliott J, Menaker M (1975c) On the role of eyes and brain photoreceptors in the sparrow: arrhythmicity in constant light. J Comp Physiol 102:263–268.

Menaker M (1965) Circadian rhythms and photoperiodism in *Passer domesticus*. In: Aschoff J (ed) Circadian Clocks. Amsterdam, North Holland, pp. 385–395.

Menaker M (1968) Light perception by extra-retinal receptors in the brain of the sparrow. Proceedings, 76th Annual Convention, APA, pp. 299–300.

Menaker M, Roberts R, Elliott J, Underwood H (1970) Extraretinal light perception in the sparrow III: The eyes do not participate in photoperiodic photoreception. Proc Natl Acad Sci USA 67:320–325.

Nyce J, Binkley S (1977) Extraretinal photoreception in chickens: entrainment of the circadian locomotor activity rhythm. Photochem Photobiol 25:529–531.

Oksche A, Kirschstein H (1969) Elektronenmikroskopische Untersuchungen am Pinealorgan von *Passer domesticus*. Z Zellforsch 102:214–241.

Ralph C, Binkley S, MacBride S, Klein D (1975) Regulation of pineal rhythms in chickens: effects of blinding, constant light, constant dark, and superior cervical ganglionectomy. Endocrinology 97:1373–1378.

Riebman J, Binkley S (1979) Regulation of pineal rhythms in chickens: N-acetyltransferase activity in homogenates. Comp Biochem Physiol 63C:291–296.

Takahashi J (1981) Neural and endocrine regulation of avian circadian systems. Ph.D. thesis, University of Oregon.

Takahashi J, Hamm H, Menaker M (1980) Circadian rhythms of melatonin release from individual superfused chicken pineal glands in vitro. Proc Natl Acad Sci USA 77:2319–2322.

Turek F, McMillan J, Menaker M (1976) Melatonin: effects on the circadian locomotor rhythm of sparrows. Science 194:1441–1443.

Underwood H, Siopes T (1984) Circadian organization in Japanese quail. J Exp Zool 232:557–566.

Wainwright S (1980) Diurnal cycles in serotonin acetyltransferase activity and cyclic GMP content of cultured chick pineal glands. Nature (London) 285:478–480.

Wainwright S, Wainwright L (1979) Chick pineal serotonin acetyltransferase: a diurnal cycle maintained in vitro and its regulation by light. Can J Biochem 57:700–709.

Wainwright S, Wainwright L (1981a) Regulation of chick pineal serotonin-N-acetyltransferase activity by light and extra-pineal factor(s). In: Pineal Function, Amsterdam, North Holland, pp. 199–210.

Wainwright S, Wainwright L (1981b) The relationship between variations in levels of serotonin acetyltransferase activity and cGMP content in cultured chick pineal glands. Can J Biochem 59:593–604.

Zimmerman N, Menaker M (1975) Neural connections of sparrow pineal: role in circadian control of activity. Science 190:477–479.

Zimmerman N, Menaker M (1979) The pineal gland: a pacemaker within the circadian system of the house sparrow. Proc Natl Acad Sci USA 76:999–1003.

Chapter 6

Daylength and Control of Seasonal Reproduction in Male Birds

FRED E. WILSON and RICHARD S. DONHAM

I. Introduction

Reproduction in most species of birds is discontinuous, or seasonal. Because young are produced during the season that favors their survival (i.e., when environmental conditions are optimal), seasonal reproduction doubtless represents an adaptation to a periodic environment. Although any source of environmental information that reliably predicts forthcoming favorable conditions may be used to trigger reproduction and/or to synchronize it with the optimum season, daylength, because of its consistent and precise relationship to season, especially at middle and high latitudes, seems particularly well suited for such purposes. The importance of daylength was demonstrated experimentally more than 60 years ago when Rowan (1925) reported that extending the photoperiod induced testicular growth and courtship song in slate-colored juncos exposed to the rigors of winter in central Alberta! Since that initial demonstration, photoperiodic control of reproduction has been documented for about sixty species of birds in fifteen families (Lofts et al. 1970, Farner 1975). All of the known photoperiodic species are "long-day" breeders. Although other sources of information (e.g., behavioral interactions, weather) may be important in the fine-tuning and eventual success of seasonal reproduction (Moore 1982, Wingfield et al. 1983, Runfeldt and Wingfield 1985, Wingfield 1985), daylength is clearly the primary proximate factor that controls the onset of the breeding season in most temperate-zone species of birds. We review here the physiological basis of the photoinduced testicular cycle, including the termination of the reproductive effort.

II. Reproductive Hormones

During the past decade, the annual cycles of reproductive hormones have been carefully described for some of the more popular photoperiodic species [e.g., white-crowned sparrows (*Zonotrichia* sp.), Wingfield and Farner 1978a,b; Japanese quail (*Coturnix coturnix japonica*), Follett and Maung 1978, Robinson and Follett 1982; European starling (*Sturnus vulgaris*), Dawson and Goldsmith 1982, 1984, Dawson 1983; and the congeners, red grouse and willow ptarmigan (*Lagopus* sp.), Stokkan and Sharp 1980a, Stokkan et al. 1985)]. There is a single period of gamete production which is accompanied by major and prolonged increases of plasma luteinizing hormone (LH), follicle-stimulating hormone (FSH), and testosterone. The major increases of LH and FSH are often coincident with testicular recrudescence in the male. Frequently, in natural populations, LH and testosterone are elevated as territories are established and mates are attracted. There is almost always a peak of LH and of testosterone coincident with nest building or initiation of a clutch by the female. As incubation begins, LH and testosterone decrease but, if a renesting attempt is required, there is often a second peak of hormones; in multiple-brooded forms, there usually is an endocrine reflection of each nesting effort. Hormone levels then decrease as postnuptial molt occurs and, in general, remain minimal through late summer, fall, and early winter. In many (if not most) photoperiodic species, plasma levels of gonadotropins and sex steroids decrease and the reproductive system collapses at a time when daylength is maximal.

Autumnal peaks in the plasma levels of LH (which frequently are not accompanied by increased concentrations of plasma testosterone) have been described in red grouse (Stokkan et al. 1985), the rook (Lincoln et al. 1980), the mallard (Haase et al. 1975, Donham 1979), and the great tit (Rohss and Silverin 1983). In these cases, autumnal peaks of LH are correlated with territorial and/or courtship behaviors.

Although luteinizing hormone-releasing hormone (LHRH) remains to be measured in avian portal plasma, two LHRH peptides have been isolated from chicken hypothalamus and characterized structurally (King and Millar 1984, Miyamoto et al. 1984). Because the activity of LHRH-I ([Gln^8]-LHRH) is severalfold less than that of LHRH-II ([His^5,Trp^7,Tyr^8]-LHRH), Millar and King (1984) speculated that LHRH-I either has a primary function unrelated to gonadotropin stimulation or acts as a modulator of gonadotropin secretion in concert with LHRH-II. However, Mikami's failure to detect LHRH-II immunocytochemically in axons that terminate in the median eminence in male quail (personal communication) argues that LHRH-I is the hypophysiotropic hormone.

Using antiserum prepared against mammalian LHRH, two populations of LHRH-immunoreactive neurons have been demonstrated in the avian hypothalamus (Bons et al. 1978, Mikami and Yamada 1984). One is in the preoptic-anterior hypothalamic region; the other, in the tuberal complex. Both populations project axons to the zona externa of the median eminence. Controversy surrounds the tuberal population, as it cannot be detected in some species (Sterling and Sharp 1982, Blähser 1984).

LHRH secreted from axon terminals in the median eminence is doubtless transported to the pituitary gland via the hypophysial portal system. As demonstrated by Hasegawa et al. (1984), both LHRH-I and mammalian LHRH stimulate dose-dependent release of LH from chicken pituitary cells in culture.

The same neuropeptides evoke the release of LH and FSH from the quail pituitary gland in vitro and in vivo (Hattori et al. 1985), and the equipotent mammalian LHRH elicits a dose-dependent increase in plasma LH concentration in white-crowned sparrows (Wingfield et al. 1979) and ducks (Bluhm 1985). Recent results on quail (Hattori et al. 1986) suggest that production and release of FSH are primarily autonomous, whereas production and release of LH are strictly controlled by LHRH.

III. Physiological Basis of Photoinduced LH Secretion

Given the neuroendocrine context just described, how do long days "turn on" seasonal reproduction? Recent research on this question has emphasized two potential mechanisms by which gonadotropin secretion may be stimulated: (1) a photoinduced decrease in the efficacy of gonadal negative feedback, and (2) an increase in direct photoperiodic drive exerted on the neuroendocrine apparatus.

In the mid-1970s, Wilson and Follett (1974, 1977) showed that castration increased the concentration of plasma LH in American tree sparrows (*Spizella arborea*) held on a short daily photoperiod. Wilson (1985a) later demonstrated that testosterone reversed the stimulatory effect of castration. Because LH concentrations in castrated tree sparrows retained on short days often exceed peak concentrations in intact photostimulated males, the possibility exists that long days stimulate LH secretion in intact males *merely* by reducing the efficacy of testosterone negative feedback. If long days stimulate LH secretion *solely* by relaxing testosterone negative feedback, then plasma LH concentrations in birds in which feedback is inoperative because of castration should be both maximal and independent of daylength. Wilson and Follett (1977) tested that prediction and found that LH concentrations in castrated tree sparrows transferred to long days always exceeded those in castrated birds kept on short days (Figure 6-1). So, do the results for tree sparrows, plus similar results for at least six other species [quail (Gibson et al. 1975, Davies et al. 1976, Urbanski and Follett 1982); white-crowned sparrows (Mattocks et al. 1976, Wingfield et al. 1980); red grouse (Sharp and Moss 1977); willow ptarmigan (Stokkan and Sharp 1980b); turkeys (El Halawani et al. 1980); and canaries (Storey and Nicholls 1981)], mean that long days stimulate LH secretion by a mechanism that is independent of testosterone negative feedback? Perhaps, but it must be emphasized that LH concentrations in castrated birds exposed to short days would also be suppressed if the hypothalamus–pituitary axis were sensitive to circulating androgens which may persist after castration. In order to evaluate the inhibitory effect of putative castration-resistant androgens, Wilson (1985a) treated castrated tree sparrows with the pure antiandrogen cyproterone (free alcohol) at a dose severalfold greater than an inhibitory dose of testosterone. The free alcohol of cyproterone reportedly has only antiandrogenic effects (Neumann et al. 1977) and should increase LH levels in the blood of short-day

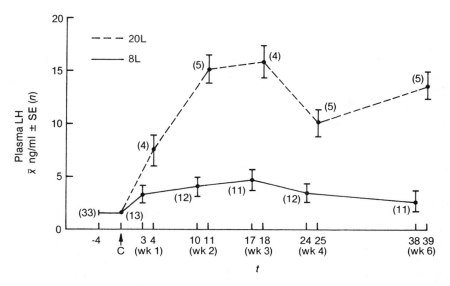

Figure 6-1. Concentrations of plasma LH in male tree sparrows castrated and transferred to a 20-h daily photoperiod (dashed line) or castrated and retained on an 8-h daily photoperiod (solid line). t, time in days; C, day of castration. 20L vs. 8L: $p < 0.05$ (ANOVA/LSD test) at weeks 1, 2, 3, 4, and 6. (From Wilson and Follett 1977.)

birds if androgens persist after castration. Because concentrations of plasma LH in both short-day and long-day castrated birds that received cyproterone never exceeded plasma LH concentrations in control birds that received vehicle, extratesticular androgens probably do not inhibit LH secretion in male tree sparrows. Therefore, differences in plasma LH concentration between short- and long-day castrated birds most likely reflect an effect of photostimulation that is independent of testosterone negative feedback. Accordingly, long days may induce LH secretion in male tree sparrows by direct neurogenic stimulation of LHRH release.

That view is supported by data of Urbanski and Follett (1982) (Figure 6-2), which show that plasma LH concentration in castrated quail is proportional to daylength. In the absence of gonadal feedback, LH concentration presumably reflects the intensity of direct neurogenic stimulation, or photoperiodic drive, imposed by daylength on the LH control mechanism. Generally among birds (Wilson and Follett 1974, 1977, Gibson et al. 1975, Davies et al. 1976, Mattocks et al. 1976, El Halawani et al. 1980, Stokkan and Sharp 1980b, Wingfield et al. 1980, Storey and Nicholls 1981, 1982, Urbanski and Follett 1982, Wilson 1985a), though not without exception (Haase et al. 1982, Dawson and Goldsmith 1984, Stokkan and Sharp 1984), plasma LH is more concentrated in photostimulated castrated birds than in nonphotostimulated castrated birds and, in addition, castrated birds show photoinduced and annual cycles in plasma LH concentration (see also Storey and Nicholls 1982). Thus, in quail and most other species,

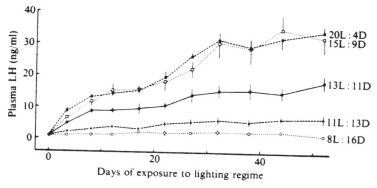

Figure 6-2. Concentrations (mean ± SEM) of plasma LH in groups of six to eleven Japanese quail which were castrated on day 0 and immediately transferred from 8-h light:16-h darkness (8L:16D) to one of the photoperiods shown. (From Urbanski and Follett 1982.)

long days likely stimulate LH secretion by increasing the photoperiodic drive imposed on the LH control mechanism.

Whether a photoinduced change in the efficacy of testosterone negative feedback *contributes* to the seasonal increase in LH secretion is an important question, but one that cannot be answered using castrated birds *without* testosterone replacement, in which plasma LH concentrations are dependent upon daylength. However, the recent development of an experimental paradigm which distinguishes testosterone feedback-dependent and -independent effects of photostimulation on LH secretion (Wilson 1985a) has permitted that question to be addressed. The paradigm (Figure 6-3) assumes that plasma LH concentration is independent of testosterone at both low and high concentrations of testosterone but a function of testosterone over an intermediate range of concentrations. It incorporates the feedback-independent effect of photostimulation, which can be quantified in castrated birds without replacement testosterone, and presumes two feedback performance characteristics: sensitivity (i.e., the change in LH per unit change in testosterone, revealed as the slope of the line relating plasma LH concentration to dose of replacement testosterone) and threshold of inhibition, or set point (i.e., the minimum dose of replacement testosterone that suppresses plasma LH concentration). Both sensitivity and set point are potentially quantifiable in castrated birds given replacement testosterone. The solid profiles in Figure 6-3 depict the putative relationship between LH and testosterone in castrated birds held on short days, while the broken profiles depict the same relationship superimposed upon the testosterone feedback-independent effect of photostimulation.

As summarized in Table 6-1, long days reduce, rather dramatically initially but more gradually thereafter, in male tree sparrows the sensitivity of the hypothalamus–pituitary axis to testosterone. Variability in plasma LH concentration at low doses of replacement testosterone has so far precluded direct

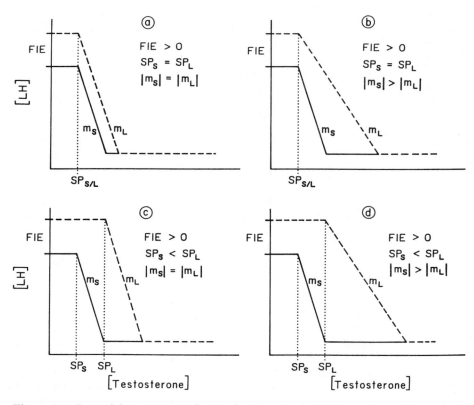

Figure 6-3. Potential testosterone feedback-dependent and -independent effects of photostimulation on plasma LH concentration in male tree sparrows. (a) Long daily photoperiods may stimulate LH secretion solely by a mechanism that operates independently of testosterone feedback (FIE), or (b–d) long daily photoperiods may stimulate LH secretion by simultaneously activating a testosterone feedback-independent mechanism and altering one or both performance characteristics [set point (SP) and/or sensitivity (m)] of the testosterone feedback mechanism. Only photoinduced increases in SP and reductions in m are illustrated. Scales are not necessarily arithmetic, and the relationship between plasma LH and testosterone at concentrations above threshold is not necessarily rectilinear. Subscript S and solid lines signify short-day responses; subscript L and broken lines, long-day responses. The dotted lines identify the minimum concentration of testosterone that suppresses plasma LH (i.e., SP); $|\,m\,|$ = absolute value of m. (From Wilson 1985a.)

measurement of the set point, but the putative set point (i.e., an *indirect* measure based on both the variance in plasma LH concentration in castrated birds without replacement testosterone and the regression of plasma LH concentration on dose of testosterone in castrated birds without replacement testosterone) is dependent upon daylength and increases significantly when daylength is increased (Wilson 1985a). Such changes in feedback performance characteristics (i.e., a reduction in sensitivity and an increase in set point) render testosterone

Table 6-1. Sensitivity and Putative Set Point of the Testosterone (T) Negative Feedback Mechanism in Castrated Tree Sparrows Held on an 8-h Daily Photoperiod (8L:16D) or Transferred to a 20-h Daily Photoperiod (20L:4D)

Photoperiod[a]	Sensitivity[b,c] (nmol T)	Putative set point[c,d] $(1/\mu mol\ T)$
8L:16D	-49[a]	2[a]
20L:4D, day 8	- 3.53[b]	70[b]
20L:4D, day 15	- 2.59[b,c]	98[b]
20L:4D, day 22	- 2.17[c]	124[b]
20L:4D, day 43	- 1.64[d]	110[b]

[a]Day signifies postcastration day. Birds were transferred to 20L:4D on day 0.
[b]Change in log plasma LH per μmol change in T.
[c]Values that share common superscript letters do not differ ($p >$ 0.05). A conservative one-sided t-test was used to compare sensitivities and an applied regression analysis to compare putative set points.
[d]Minimum concentration of T that suppresses plasma LH.
(From Wilson 1985a.)

a less potent inhibitor of LH secretion. Thus, long days stimulate LH secretion in male tree sparrows not only by increasing photoperiodic drive, but also by reducing the efficacy of testosterone negative feedback.

Several investigators [among them Gogan (1970), Cusick and Wilson (1972), Davies et al. (1976), Sharp and Moss (1977), and Stokkan and Sharp (1980b)], all without benefit of an experimental design that distinguished real from apparent changes in feedback sensitivity, have speculated that sensitivity of the hypothalamus–pituitary axis to testosterone varies during an annual testicular cycle. Confirmation of such speculation in tree sparrows (Wilson 1985a, 1986) gives the hypothesis needed credibility, but it seems unlikely that changes in testosterone feedback sensitivity play a fundamental role in timing the onset of seasonal reproduction in tree sparrows or in other species in which LHRH appears to be secreted mainly in response to direct neurogenic stimulation. On the other hand, plasma LH concentration is both elevated and independent of daylength in photosensitive castrated starlings (Dawson and Goldsmith 1984, Goldsmith and Nicholls 1984a, Nicholls et al. 1984), and plasma LH concentration remains elevated throughout the year in castrated wild mallards exposed to natural daylengths (Haase et al. 1982). Changes in sensitivity of the hypothalamus–pituitary axis to steroid feedback may, therefore, be important in modulating LH secretion in intact starlings and mallards exposed to different daylengths. However, as noted by Dawson and Goldsmith (1984), in the absence of gonadal steroids, the hypothalamus–pituitary unit of the starling may produce LH at its maximum rate even on short days, thus obscuring any direct effect of further photoperiodic stimulation. Measurements of hypothalamic LHRH content in both photostimulated and nonphotostimulated castrated birds may clarify this issue, especially if LHRH is a more sensitive indicator of photoperiodic drive than is plasma LH.

Konishi et al. (1985) recently proposed that a retinal system modulates hypothalamic sensitivity to testosterone negative feedback in quail. When confirmed, it will be important to establish whether the influence of the eyes is mediated hormonally, possibly via melatonin, or neurally via a retinohypothalamic pathway. How long days reduce the sensitivity of the hypothalamus–pituitary axis to testosterone negative feedback is unknown, but changes in the concentration and/or affinity of receptors for testosterone or LHRH, or changes in 5α- or 5β-reductase, aromatase, and/or LHRH peptidase activities, possibly are involved.

IV. Physiological Basis of Photorefractoriness

It is well known that chronically photostimulated birds of numerous species eventually become refractory, or insensitive, to the gonadostimulatory effect of long days. Photorefractoriness may be either absolute (i.e., daylengths longer than that which caused refractoriness do not induce gonadotropin secretion or testicular recrudescence) or relative (i.e., daylengths longer than that which caused refractoriness do induce positive pituitary and testicular responses). The natural consequence of photorefractoriness is termination of reproduction, which, for species that live in a periodic environment, is obviously adaptive.

In 1949, Burger considered photorefractoriness "the greatest unsolved problem in (avian) reproduction." Apart from the currently debated question of whether photorefractoriness evolved independently on a number of separate occasions (Farner 1964) or whether the mechanism of photorefractoriness has been highly conserved during avian evolution (Follett and Nicholls 1985), the physiological basis of photorefractoriness has not yet been clearly established for any species.

Among the hypotheses that have been advanced to explain the etiology of photorefractoriness (for a review, see Robinson and Follett 1982), two have attracted considerable interest, especially recently. The more controversial hypothesis dates to the 1960s and proposes that photorefractoriness is triggered by a photoinduced hypersensitivity of the hypothalamus–pituitary axis to androgen. The availability of the castration/testosterone-replacement paradigm described earlier has permitted a direct test of that hypothesis. In a recent study (Wilson 1986), changes in photoperiodic drive and in sensitivity to testosterone negative feedback were monitored in male tree sparrows during the transition from photosensitivity to photorefractoriness. The results, which are summarized in Table 6-2, indicate that a testosterone-independent reduction in net photoperiodic drive between days 43 and 50 of photostimulation preceded any change in sensitivity to testosterone. Subsequently, sensitivity did not increase, as predicted by the hypothesis, but decreased as photoperiodic drive was reduced even further. Thus, in male tree sparrows, photorefractoriness is triggered by a testosterone-independent, but otherwise unidentified, mechanism.

Table 6-2. Net Photoperiodic Drive and Sensitivity of the
Testosterone (T) Negative Feedback Mechanism in Chronically
Photostimulated Male Tree Sparrows

Days on 20L:4D	Net photoperiodic drive[a] $[\bar{x}$ μg LH/l $\pm$ SEM $(n)]$	Sensitivity[b] $(1/\mu mol$ T)
36	$27.34^{a} \pm 1.93$ (7)	-2.18^{a}
43	$32.74^{a} \pm 4.42$ (7)	-2.18^{a}
50	$19.99^{b} \pm 2.55$ (7)	-1.81^{a}
57	$8.94^{c} \pm 0.71$ (7)	-1.14^{b}

[a]Data from castrated tree sparrows without testosterone replacement.
Means that share common superscript letters are not significantly different ($p > 0.05$; LSD multiple comparison test after ANOVA with
repeated measures).
[b]Change in log plasma LH per μmol change in replacement testosterone
(≤ 0.49 μmol). Values that share a common superscript letter are not
significantly different ($p > 0.05$; t-test for differences in slope).
(Modified from Wilson 1986.)

A similar interpretation may be advanced for other passerine species, for
castrated birds of such species become photorefractory (Mattocks et al. 1976,
Storey et al. 1980, Storey and Nicholls 1982, Dawson and Goldsmith 1984), as
do intact and castrated birds in which cycles of photoinduced gonadotropin
secretion have been abolished by exogenous testosterone (Storey and Nicholls
1981; see also Stokkan and Sharp 1980c).

Castration does not elevate plasma LH concentration in photorefractory tree
sparrows (Wilson and Follett 1974, Wilson 1985b), canaries (Nicholls and Storey
1976), starlings (Dawson and Goldsmith 1984, Goldsmith and Nicholls 1984b),
or white-crowned sparrows (Mattocks et al. 1976, Wingfield et al. 1980, Mattocks
1985). Moreover, the antiandrogens cyproterone and flutamide fail to elevate,
and replacement testosterone fails to suppress, the concentration of plasma LH
in castrated, photorefractory tree sparrows (Wilson 1985b). Taken together,
the data on passerine species suggest that a testosterone-independent mechanism
also maintains photorefractoriness.

On the contrary, castration evokes an LH response in photorefractory red
grouse (Sharp and Moss 1977) and wild mallards (Haase et al. 1982) and, as
the data in Figure 6-4 show, in photorefractory willow ptarmigan exposed to
continuous light (Stokkan and Sharp 1980b, 1984). Moreover, chronically castrated ptarmigan and mallards held on natural daylengths do not show the characteristic reduction in plasma LH concentration that marks the onset of photorefractoriness in intact males. Such data suggest that the low levels of plasma
LH typical of photorefractory grouse, ptarmigan, and mallards are maintained
by gonadal negative feedback. The low concentration of plasma testosterone
in photorefractory ptarmigan suggests further that the hypothalamus–pituitary
axis may be highly sensitive to testosterone negative feedback. However, because the LH response to castration is greater in photosensitive than in pho-

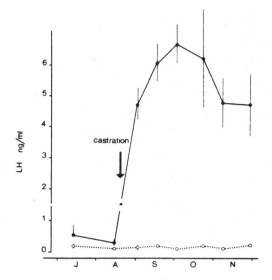

Figure 6-4. Changes in concentrations of plasma LH in castrated (solid line, $n = 6$) and intact (dotted line, $n = 6$) photorefractory willow ptarmigan exposed to continuous light. Vertical bars indicate SEM. When SEM is not shown, it is covered by the point symbol. (From Stokkan and Sharp 1980b.)

torefractory ptarmigan, the mechanism of photorefractoriness in ptarmigan likely also includes a testosterone-independent component.

Species that maintain photorefractoriness primarily by testis-dependent or -independent mechanisms may differ mainly in the balance between photoperiodic drive and androgen negative feedback (Wilson 1985b). In species with a testis-dependent mechanism, androgen feedback overrides photoperiodic drive, in which case castration will induce LH secretion. On the other hand, if photorefractoriness is maintained by a testis-independent mechanism that overrides photoperiodic drive, eliminating androgen feedback by castration will not augment LH secretion nor will androgen replacement suppress it.

Another hypothesis that has been advanced to explain photorefractoriness, and currently the most promising, is that photorefractoriness is the culmination of a thyroid-dependent inhibitory process that develops gradually and progressively from the time that birds first experience long days (Goldsmith and Nicholls 1984c, Nicholls et al. 1984).

As recently confirmed by Goldsmith and Nicholls (1984d) (Figure 6-5), thyroidectomy performed before, but not after, photosensitive starlings are exposed to long days prevents the spontaneous testicular regression normally caused by chronic photostimulation (Wieselthier and van Tienhoven 1972). Thyroidectomy also prevents the photoinduced increase in plasma prolactin concentration (Figure 6-5) which Ebling et al. (1982) and others (Dawson and Goldsmith 1983, 1984, Goldsmith and Nicholls 1984b) have established is correlated with spontaneous testicular regression in thyroid-intact starlings. Goldsmith and Nicholls (1984d) showed that castration of thyroidectomized starlings, which had been held for about a year on long days, markedly elevated plasma gonadotropin concentrations, whereas castration of photorefractory thyroid-intact

Figure 6-5. Testicular width and plasma concentrations of prolactin (mean ± SEM) in intact (dashed line) and thyroidectomized (solid line) male starlings exposed to long (18-h) daily photoperiods. The thyroidectomized birds did not show a rise in prolactin and did not become photorefractory. (From Nicholls et al. 1984.)

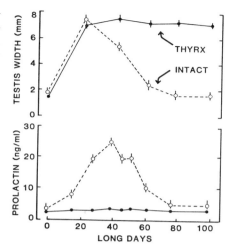

starlings did not. The castration-induced elevation in thyroidectomized birds, which is uncharacteristic of photorefractory passerines, including starlings, verifies that thyroidectomy prevented photorefractoriness. Because hypothalamic LHRH content is significantly lower in thyroid-intact photorefractory starlings than in thyroidectomized photosensitive starlings (Dawson et al. 1985a), photorefractoriness in starlings may be due to a thyroid-mediated deficiency of LHRH.

Goldsmith et al. (1985) demonstrated that thyroxine, administered in drinking water for 14 days, induced testicular regression and hyperprolactinemia in chronically photostimulated, thyroidectomized starlings. Thyroidectomized controls denied thyroxine replacement maintained testes of maximum size and low plasma concentrations of prolactin. Such data provide additional evidence that a thyroid-dependent process triggers photorefractoriness.

As first shown by Schwab (1970) and confirmed by Hamner (1971) and Dawson et al. (1985b), starlings held on an 11-h daily photoperiod do not become photorefractory but maintain mature testes indefinitely. As shown by Goldsmith and Nicholls (1984c), thyroxine, administered in drinking water for 5 weeks, induced testicular regression and raised plasma prolactin concentration in thyroid-intact starlings held on an 11-h daily photoperiod. The inhibitory effect of thyroxine on the testes is not likely direct, as exogenous thyroxine also blocked pituitary gonadotropin secretion in castrated starlings held on an 11-h daily photoperiod. However, thyroxine-treated intact starlings did become photorefractory, as raising the daylength to 18 h did not induce testicular recrudescence. As expected, when thyroid-intact starlings held on an 11-h daily photoperiod were denied exogenous thyroxine and later transferred to an 18-h daily photoperiod, photorefractoriness was induced, as manifested by testicular regression and hyperprolactinemia. High doses of mammalian prolactin administered to thyroid-intact starlings held on an 11-h daily photoperiod also induced testicular regression. However, because the testes promptly recrudesced when prolactin treatment ceased and daylength increased, exogenous prolactin did not induce a photorefractory state.

Dawson (1984) has measured plasma thyroxine, but not plasma triiodothyronine, concentrations in starlings exposed to different daylengths. As noted before, photosensitive male starlings transferred to an 11-h daily photoperiod become sexually mature, but not photorefractory. Plasma thyroxine levels in such birds are rather stable, as they are in sexually immature starlings held on an 8-h daily photoperiod. However, in males held either on an 18-h daily photoperiod or on a 13-h daily photoperiod (two photoregimes that induce photorefractoriness), plasma thyroxine concentrations peaked after 2 to 3 weeks. Dawson (1984) concluded that increased thyroxine secretion is associated with photorefractoriness rather than with testicular growth. The generality of this scheme is presently uncertain, but plasma profiles of thyroid hormones in white-crowned sparrows and house sparrows (*Passer domesticus*) on natural photoperiods (Smith 1982) and in willow ptarmigan after photostimulation (Klandorf et al. 1982) suggest that increasing or long photoperiods induce, albeit with some delay, an elevation in plasma thyroxine concentration. As Goldsmith and Nicholls (1984c) point out, peak thyroxine levels may not necessarily correlate with the actual time of testicular regression, which is simply the culmination of a series of events initiated earlier.

Follett and Maung (1978) presented data showing changes in testicular mass, area of the androgen-dependent cloacal gland, and concentrations of plasma gonadotropins and testosterone in quail exposed to natural daylengths in Great Britain. As suggested by their data and confirmed by Robinson and Follett (1982), daylengths that are gonadostimulatory in spring are not gonadostimulatory in autumn, and so quail also become photorefractory. However, quite unlike species that exhibit absolute photorefractoriness, refractory quail will still respond to a longer daylength. Thus, quail are said to exhibit relative photorefractoriness. Follett and Nicholls (1984, 1985) have questioned whether absolute and relative photorefractoriness are separate physiological processes or represent the extremes of a single inhibitory process that is dependent both upon long days and upon the thyroid gland. Their evidence indicates that testicular regression triggered by reduced daylength is the manifestation of a thyroid-dependent process initiated during prior exposure to a long daily photoperiod.

In contrast to the observations of Underwood (1975) on house sparrows, the observations of Homma et al. (1972) and Siopes and Wilson (1980) on quail and those of Yokoyama and Farner (1976) on white-crowned sparrows suggest a gonadoinhibitory role for the eyes. As demonstrated most recently by Konishi et al. (1985), testicular regression induced by transferring sighted quail from a 14-h daily photoperiod to an 8-h daily photoperiod is blocked by bilateral optic enucleation. This observation suggests that the eyes may be components of the mechanism that terminates reproduction when quail are exposed to short days in the laboratory or to decreasing daylengths after the summer solstice. The role of the eyes in the mechanism of relative photorefractoriness and interactions among daylength, the eyes, the thyroid gland, and prolactin in species that exhibit absolute photorefractoriness are issues yet to be explored.

V. Physiological Basis of Photosensitivity

In nature, the hiatus in reproduction initially caused by absolute photorefractoriness and later maintained by the short days of autumn and winter is terminated only when photosensitivity is restored and daylength again exceeds a critical duration. It is well known that exposure of photorefractory birds to short days dissipates photorefractoriness, but how is photosensitivity restored? That question cannot yet be answered satisfactorily, but some recent observations are pertinent to it.

Although probably unrelated causally, gonadal feedback inhibition is reinstated when photosensitivity is restored in canaries (Nicholls and Storey 1976), willow ptarmigan (Stokkan and Sharp 1980b), and starlings (Nicholls et al. 1984). As shown by Stokkan and Sharp (1980b), photorefractory ptarmigan, castrated and transferred from continuous light to a short daily photoperiod, show no immediate postcastration LH response, although a rather robust response is evident after several weeks on short days. Over the same interval, LH concentrations of intact birds remain low. The latency of the postcastration increase in LH concentration suggests that photoperiodic drive is disengaged until photosensitivity is restored, which implies, in turn, that photorefractoriness stems from a defect in the neural processing of photoperiodic information.

As suggested by Nicholls et al. (1984), an increase in hypothalamic LHRH content likely accompanies the recovery of photosensitivity in starlings retained on short days, as photorefractory starlings sampled after 12 weeks on an 18-h daily photoperiod have significantly less LHRH than do photosensitive starlings sampled after protracted exposure to an 8-h daily photoperiod. Thus, restoration of photosensitivity may be linked to a renewed ability or augmented capacity to synthesize LHRH.

The recent observations of Dawson et al. (1985c, 1987) that thyroidectomy terminates photorefractoriness both in adult and in first-year starlings retained on long days suggest that photorefractoriness is maintained by thyroid hormones and that photosensitivity may be restored by reversing the processes that induced photorefractoriness. If, as these data suggest, thyroidectomy initiates a process similar to that initiated by short days, then exogenous thyroxine might block recovery of photosensitivity in photorefractory starlings transferred to short days. If the eyes are shown to play a role in the development of photorefractoriness, it may be productive to explore interactions among the eyes and the thyroid gland during the transition from photorefractoriness to photosensitivity.

VI. Summary

Research of recent years has culminated in rather thorough descriptions of the seasonal endocrinology of several photoperiodic species of birds. In addition to these advances, there has been progress in our understanding of the mechanisms by which avian reproduction is turned on and subsequently terminated.

The foregoing discussion has focused, albeit with considerable selectivity, on recent advances related to three pivotal events upon which the strategy of seasonal breeding depends in many photoperiodic species of birds, namely, long day-mediated activation of the hypothalamus–pituitary–testis unit; its deactivation by the long day-mediated process of absolute photorefractoriness; and short day-mediated restoration of photosensitivity. Long photoperiods appear to initiate reproduction by one or both of two mechanisms: (1) an androgen feedback-independent effect of long days, which results in direct photoperiodic drive of gonadotropin secretion in proportion to the length of the photoperiod, and (2) a photoinduced reduction in the efficacy of androgen-dependent negative feedback. The literature suggests that there may be species-dependent differences in the relative roles of the two mechanisms, but the weight of evidence favors the former.

Investigations of the decades-old enigma of photorefractoriness have produced data not currently resolvable into a coherent theory, but which point the way for future investigations. The proposal that refractoriness is initiated by a photoinduced hypersensitivity of the hypothalamus–pituitary axis to androgen is supported by data from some waterfowl and galliforms. However, in several passeriforms, data suggest that a testosterone-independent mechanism initiates and maintains photorefractoriness. In addition, a systematic series of investigations has also revealed that absolute photorefractoriness in starlings, and probably relative photorefractoriness in quail, is a thyroid-dependent process induced by exposure to long daylengths. The importance of this hypothesis, if upheld for other species, is difficult to overestimate.

Recent observations show that restoration of photosensitivity in photorefractory birds exposed to short days is correlated with an increase in hypothalamic LHRH content, concomitant with the reappearance of an LH response to castration. Both of these phenomena appear to be part of a thyroid-mediated process. Clearly, this is a time of significant advances in our understanding of photoperiodic phenomena in birds.

References

Blähser S (1984) Peptidergic pathways in the avian brain. J Exp Zool 232:397–403.

Bluhm CK (1985) Seasonal variation in pituitary responsiveness to luteinizing hormone-releasing hormone of mallards and canvasbacks. Gen Comp Endocrinol 58:491–497.

Bons N, Kerdelhué B, Assenmacher I (1978) Mise en évidence d'un deuxième système neurosécrétoire à LH-RH dans l'hypothalamus du canard. C R Acad Sci Paris Ser D 287:145–148.

Burger JW (1949) A review of experimental investigations on seasonal reproduction in birds. Wilson Bull 61:211–230.

Cusick EK, Wilson FE (1972) On control of spontaneous testicular regression in tree sparrows (*Spizella arborea*). Gen Comp Endocrinol 19:441–456.

Davies DT, Goulden LP, Follett BK, Brown NL (1976) Testosterone feedback on luteinizing hormone (LH) secretion during a photoperiodically induced breeding cycle in Japanese quail. Gen Comp Endocrinol 30:477–486.

Dawson A (1983) Plasma gonadal steroid levels in wild starlings (*Sturnus vulgaris*) during the annual cycle and in relation to the stage of breeding. Gen Comp Endocrinol 49:286–294.

Dawson A (1984) Changes in plasma thyroxine concentrations in male and female starlings (*Sturnus vulgaris*) during a photo-induced gonadal cycle. Gen Comp Endocrinol 56:193–197.

Dawson A, Goldsmith AR (1982) Prolactin and gonadotrophin secretion in wild starlings (*Sturnus vulgaris*) during the annual cycle and in relation to nesting, incubation, and rearing young. Gen Comp Endocrinol 48:213–221.

Dawson A, Goldsmith AR (1983) Plasma prolactin and gonadotrophins during gonadal development and the onset of photorefractoriness in male and female starlings (*Sturnus vulgaris*) on artificial photoperiods. J Endocrinol 97:253–260.

Dawson A, Goldsmith AR (1984) Effects of gonadectomy on seasonal changes in plasma LH and prolactin concentrations in male and female starlings (*Sturnus vulgaris*). J Endocrinol 100:213–218.

Dawson A, Follett BK, Goldsmith AR, Nicholls TJ (1985a) Hypothalamic gonadotrophin-releasing hormone and pituitary and plasma FSH and prolactin during photostimulation and photorefractoriness in intact and thyroidectomized starlings (*Sturnus vulgaris*). J Endocrinol 105:71–77.

Dawson A, Goldsmith AR, Nicholls TJ (1985b) Development of photorefractoriness in intact and castrated male starlings (*Sturnus vulgaris*) exposed to different periods of long-day lengths. Physiol Zool 58:253–261.

Dawson A, Goldsmith AR, Nicholls TJ (1985c) Thyroidectomy results in termination of photorefractoriness in starlings (*Sturnus vulgaris*) kept in long daylengths. J Reprod Fert 74:527–533.

Dawson A, Williams TD, Nicholls TJ (1987) Thyroidectomy of nestling starlings appears to cause neotenous sexual maturation. J Endocrinol 112:R5–R6.

Donham RS (1979) Annual cycle of plasma luteinizing hormone and sex hormones in male and female mallards (*Anas platyrhynchos*). Biol Reprod 21:1273–1285.

Ebling FJP, Goldsmith AR, Follett BK (1982) Plasma prolactin and luteinizing hormone during photoperiodically induced testicular growth and regression in starlings (*Sturnus vulgaris*). Gen Comp Endocrinol 48:485–490.

El Halawani ME, Burke WH, Ogren LA, Millam JR (1980) Developmental changes in the photosensitivity of male turkeys. Gen Comp Endocrinol 40:226–231.

Farner DS (1964) The photoperiodic control of reproductive cycles in birds. Am Sci 52:137–156.

Farner DS (1975) Photoperiodic controls in the secretion of gonadotropins in birds. Am Zool 15 (Suppl 1):117–135.

Follett BK, Maung SL (1978) Rate of testicular maturation, in relation to gonadotrophin and testosterone levels, in quail exposed to various artificial photoperiods and to natural daylengths. J Endocrinol 78:267–280.

Follett BK, Nicholls TJ (1984) Photorefractoriness in Japanese quail: possible involvement of the thyroid gland. J Exp Zool 232:573–580.

Follett BK, Nicholls TJ (1985) Influences of thyroidectomy and thyroxine replacement on photoperiodically controlled reproduction in quail. J Endocrinol 107:211–221.

Gibson WR, Follett BK, Gledhill (1975) Plasma levels of luteinizing hormone in gonadectomized Japanese quail exposed to short or to long daylengths. J Endocrinol 64:87–101.

Gogan F (1970) Rétroaction des stéroides sexuels sur les fonctions gonadotropes de l'oiseau. In: Benoit J, Kordon C (eds) Neuroendocrinologie. Editions du Centre National de la Recherche Scientifique, Paris, pp. 351–362.

Goldsmith AR, Nicholls TJ (1984a) Recovery of photosensitivity in photorefractory starlings is not prevented by testosterone treatment. Gen Comp Endocrinol 56:210–217.

Goldsmith AR, Nicholls TJ (1984b) Prolactin is associated with the development of photorefractoriness in intact, castrated, and testosterone-implanted starlings. Gen Comp Endocrinol 54:247–255.

Goldsmith AR, Nicholls TJ (1984c) Thyroxine induces photorefractoriness and stimulates prolactin secretion in European starlings (*Sturnus vulgaris*). J Endocrinol 101:R1–R3.

Goldsmith AR, Nicholls TJ (1984d) Thyroidectomy prevents the development of photorefractoriness and the associated rise in plasma prolactin in starlings. Gen Comp Endocrinol 54:256–263.

Goldsmith AR, Nicholls TJ, Plowman G (1985) Thyroxine treatment facilitates prolactin secretion and induces a state of photorefractoriness in thyroidectomized starlings. J Endocrinol 104:99–103.

Haase E, Sharp PJ, Paulke E (1975) Annual cycle of plasma luteinizing hormone concentrations in wild mallard drakes. J Exp Zool 194:553–558.

Haase E, Sharp PJ, Paulke E (1982) The effects of castration on the seasonal pattern of plasma LH concentrations in wild mallard drakes. Gen Comp Endocrinol 46:113–115.

Hamner WM (1971) On seeking an alternative to the endogenous reproductive rhythm hypothesis in birds. In: Menaker M (ed) Biochronometry. National Academy of Sciences, Washington, D.C., pp. 448–461.

Hasegawa Y, Miyamoto K, Igarashi M, Chino N, Sakakibara S (1984) Biological properties of chicken luteinizing hormone-releasing hormone: gonadotropin release from rat and chicken cultured anterior pituitary cells and radioligand analysis. Endocrinology 114:1441–1447.

Hattori A, Ishii S, Wada M, Miyamoto K, Hasegawa Y, Igarashi M, Sakakibara S (1985) Effects of chicken (Gln[8])- and mammalian (Arg[8])-luteinizing hormone-releasing hormones on the release of gonadotrophins *in vitro* and *in vivo* from the adenohypophysis of Japanese quail. Gen Comp Endocrinol 59:155–161.

Hattori A, Ishii S, Wada M (1986) Different mechanisms controlling FSH and LH release in Japanese quail (*Coturnix coturnix japonica*): evidence for an inherently spontaneous release and production of FSH. J Endocrinol 108:239–245.

Homma K, Wilson WO, Siopes TD (1972) Eyes have a role in photoperiodic control of sexual activity of *Coturnix*. Science 178:421–423.

King JA, Millar RP (1984) Isolation and structural characterization of chicken hypothalamic luteinizing hormone releasing hormone. J Exp Zool 232:419–423.

Klandorf H, Stokkan KA, Sharp PJ (1982) Plasma thyroxine and triiodothyronine levels during the development of photorefractoriness in willow ptarmigan (*Lagopus lagopus lagopus*) exposed to different photoperiods. Gen Comp Endocrinol 47:64–69.

Konishi H, Wada M, Homma K (1985) Retinal modulation of the hypothalamic sensitivity to testosterone feedback in photoperiodism of quail. Gen Comp Endocrinol 59:343–349.

Lincoln GA, Racey PA, Sharp PJ, Klandorf H (1980) Endocrine changes associated with spring and autumn sexuality of the rook, *Corvus frugilegus*. J Zool (Lond) 190:137–153.

Lofts B, Follett BK, Murton RK (1970) Temporal changes in the pituitary-gonadal axis. In: Benson GK, Phillips JG (eds) Hormones and the Environment. Cambridge University Press, London, pp. 545–575 (Memoirs of the Society for Endocrinology, no. 18).

Mattocks PW Jr (1985) Absence of gonadal-hormone feedback on luteinizing hormone in both photorefractory and photosensitive white-crowned sparrows (*Zonotrichia leucophrys gambelii*). Gen Comp Endocrinol 60:455–462.

Mattocks PW Jr, Farner DS, Follett BK (1976) The annual cycle in luteinizing hormone in the plasma of intact and castrated white-crowned sparrows, *Zonotrichia leucophrys gambelii*. Gen Comp Endocrinol 30:156–161.

Mikami S-I, Yamada S (1984) Immunohistochemistry of the hypothalamic neuropeptides and anterior pituitary cells in the Japanese quail. J Exp Zool 232:405–417.

Millar RP, King JA (1984) Structure-activity relations of LHRH in birds. J Exp Zool 232:425–430.

Miyamoto K, Hasegawa Y, Nomura M, Igarashi M, Kangawa K, Matsuo H (1984) Identification of the second gonadotropin-releasing hormone in chicken hypothalamus: evidence that gonadotropin secretion is probably controlled by two distinct gonadotropin-releasing hormones in avian species. Proc Natl Acad Sci USA 81:3874–3878.

Moore MC (1982) Hormonal responses of free-living male white-crowned sparrows to experimental manipulation of female sexual behavior. Horm Behav 16:323–329.

Neumann F, Graf KJ, Hasan SH, Schenck B, Steinbeck H (1977) Central actions of antiandrogens. In: Martini L, Motta M (eds) Androgens and Antiandrogens. Raven Press, New York, pp. 163–177.

Nicholls TJ, Storey CR (1976) The effects of castration on plasma LH levels in photosensitive and photorefractory canaries (*Serinus canarius*). Gen Comp Endocrinol 29:170–174.

Nicholls TJ, Goldsmith AR, Dawson A (1984) Photorefractoriness in European starlings: associated hypothalamic changes and the involvement of thyroid hormones and prolactin. J Exp Zool 232:567–572.

Robinson JE, Follett BK (1982) Photoperiodism in Japanese quail: the termination of seasonal breeding by photorefractoriness. Proc R Soc Lond B 215:95–116.

Rohss M, Silverin B (1983) Seasonal variations in the ultrastructure of the Leydig cells and plasma levels of luteinizing hormone and steroid hormone in juvenile and adult male great tits *Parus major*. Ornis Scand 14:202–212.

Rowan W (1925) Relation of light to bird migration and developmental changes. Nature (London) 115:494–495.

Runfeldt S, Wingfield JC (1985) Experimentally prolonged sexual activity in female sparrows delays termination of reproductive activity in their untreated mates. Anim Behav 33:403–410.

Schwab RG (1970) Light-induced prolongation of spermatogenesis in the European starling, *Sturnus vulgaris*. Condor 72:466–470.

Sharp PJ, Moss R (1977) The effects of castration on concentrations of luteinizing hormone in the plasma of photorefractory red grouse (*Lagopus lagopus scoticus*). Gen Comp Endocrinol 32:289–293.

Siopes TD, Wilson WO (1980) Participation of the eyes in the photosexual response of Japanese quail (*Coturnix coturnix japonica*). Biol Reprod 23:352–357.

Smith JP (1982) Changes in blood levels of thyroid hormones in two species of passerine birds. Condor 84:160–167.

Sterling RJ, Sharp PJ (1982) The localisation of LH-RH neurones in the diencephalon of the domestic hen. Cell Tiss Res 222:283–298.

Stokkan KA, Sharp PJ (1980a) Seasonal changes in the concentrations of plasma luteinizing hormone and testosterone in willow ptarmigan (*Lagopus lagopus lagopus*) with observations on the effects of permanent short days. Gen Comp Endocrinol 40:109–115.

Stokkan KA, Sharp PJ (1980b) The roles of day length and the testes in the regulation of plasma LH levels in photosensitive and photorefractory willow ptarmigan (*Lagopus lagopus lagopus*). Gen Comp Endocrinol 41:520–526.

Stokkan KA, Sharp PJ (1980c) The development of photorefractoriness in willow ptarmigan (*Lagopus lagopus lagopus*) after the suppression of photoinduced LH release with implants of testosterone. Gen Comp Endocrinol 41:527–530.

Stokkan KA, Sharp PJ (1984) The development of photorefractoriness in castrated willow ptarmigan (*Lagopus lagopus lagopus*). Gen Comp Endocrinol 54:402–408.

Stokkan KA, Sharp PJ, Moss R (1985) Seasonal breeding in grouse and ptarmigan. In: Lofts B, Holmes WN (eds) Current Trends in Comparative Endocrinology. Hong Kong University Press, Hong Kong, pp. 711–713.

Storey CR, Nicholls TJ (1981) The effect of testosterone upon a photoperiodically induced cycle of gonadotrophin secretion in castrated canaries, *Serinus canarius*. Gen Comp Endocrinol 43:527–531.

Storey CR, Nicholls TJ (1982) A photoperiodically induced cycle of gonadotrophin secretion in intact, hemi- and fully castrated male bullfinches *Pyrrhula pyrrhula*. Ibis 124:55–60.

Storey CR, Nicholls TJ, Follett BK (1980) Castration accelerates the rate of onset of photorefractoriness in the canary (*Serinus canarius*). Gen Comp Endocrinol 42:315–319.

Underwood H (1975) Retinally perceived photoperiod does not influence subsequent testicular regression in house sparrows. Gen Comp Endocrinol 27:475–478.

Underwood H, Siopes T (1984) Circadian organization in Japanese quail. J Exp Zool 232:557–566.

Urbanski HF, Follett BK (1982) Photoperiodic modulation of gonadotrophin secretion in castrated Japanese quail. J Endocrinol 92:73–83.

Wieselthier AS, van Tienhoven A (1972) The effect of thyroidectomy on testicular size and on the photorefractory period in the starling (*Sturnus vulgaris* L.). J Exp Zool 179:331–338.

Wilson FE (1985a) Androgen feedback-dependent and -independent control of photoinduced LH secretion in male tree sparrows (*Spizella arborea*). J Endocrinol 105:141–152.

Wilson FE (1985b) An androgen-independent mechanism maintains photorefractoriness in male tree sparrows (*Spizella arborea*). J Endocrinol 107:137–143.

Wilson FE (1986) A testosterone-independent reduction in net photoperiodic drive triggers photorefractoriness in male tree sparrows (*Spizella arborea*). J Endocrinol 109:133–137.

Wilson FE, Follett BK (1974) Plasma and pituitary luteinizing hormone in intact and castrated tree sparrows (*Spizella arborea*) during a photoinduced gonadal cycle. Gen Comp Endocrinol 23:82–93.

Wilson FE, Follett BK (1977) Testicular inhibition of gonadotropin secretion in photosensitive tree sparrows (*Spizella arborea*) exposed to a winter-like day length. Gen Comp Endocrinol 32:440–445.

Wingfield JC (1985) Influences of weather on reproductive function in male song sparrows, *Melospiza melodia*. J Zool Lond (A) 205:525–544.

Wingfield JC, Farner DS (1978a) The annual cycle of plasma irLH and steroid hormones in feral populations of the white-crowned sparrow, *Zonotrichia leucophrys gambelii*. Biol Reprod 19:1046–1056.

Wingfield JC, Farner DS (1978b) Reproductive endocrinology of the white-crowned sparrow (*Zonotrichia leucophrys pugetensis*). Physiol Zool 51:188–205.

Wingfield JC, Crim JW, Mattocks PW Jr, Farner DS (1979) Responses of photosensitive and photorefractory male white-crowned sparrows (*Zonotrichia leucophrys gambelii*) to synthetic mammalian luteinizing hormone releasing hormone (Syn-LHRH). Biol Reprod 21:801–806.

Wingfield JC, Follett BK, Matt KS, Farner DS (1980) Effect of day length on plasma FSH and LH in castrated and intact white-crowned sparrows. Gen Comp Endocrinol 42:464–470.

Wingfield JC, Moore MC, Farner DS (1983) Endocrine responses to inclement weather in naturally breeding populations of white-crowned sparrows (*Zonotrichia leucophrys pugetensis*). Auk 100:56–62.

Yokoyama K, Farner DS (1976) Photoperiodic responses in bilaterally enucleated female white-crowned sparrows, *Zonotrichia leucophrys gambelii*. Gen Comp Endocrinol 30:528–533.

Chapter 7

Changes in Reproductive Function of Free-Living Birds in Direct Response to Environmental Perturbations

JOHN C. WINGFIELD

I. Introduction

There is a general concensus of opinion that onset of breeding in birds occurs when trophic resources are favorable for production and survival of young (e.g., Lack 1968, Perrins 1970). The environmental factors that provide proximate cues for development of reproductive function and timing of the onset of breeding have received considerable attention over the past 50 years (e.g., Farner and Gwinner 1980, Farner 1985, Wingfield 1980, 1983). However, the hormonal mechanisms underlying development of the gonads and associated structures, onset of breeding, transitions from sexual to parental behavior, and repetition of this temporal pattern in multiple broods have, of necessity, assumed ideal environmental conditions. Such an approach is crucial if experimental paradigms are to be controlled adequately and the mechanisms investigated. Nevertheless, free-living birds rarely experience "ideal" conditions for prolonged periods and are exposed to a variety of unfavorable environmental factors that have profound influences on reproductive success. Thus one can pose a question, "what happens when something goes awry?". For example, predators may consume the eggs or young, or storms may reduce food resources resulting in abandonment of the nest. The effects of extreme weather conditions on reproductive function are varied and appear to depend upon sex, body condition, and stage in the reproductive cycle (Gessamen and Worthen 1982, Elkins 1983, Wingfield 1984a, Wingfield et al., 1983). Very severe storms can result in significant mortality of adult birds at any time of year, especially if accompanied by accumulations of snow and ice that cover food resources (Whitmore et al. 1977, Ojanen 1979, Pullainen 1977, Tompa 1971). In other cases, storms early in spring do not stress adult birds unduly but may delay onset of breeding by retarding gonadal development or disrupting ovulation (e.g., Nice 1937, Riddle and Honeywell 1924, Marshall 1949, Green et al. 1977, Wingfield 1984a). In other words, early storms act as inhibitory supplementary information, thus adjusting onset of breeding until environmental conditions are favorable.

During the nesting phase, birds are particularly susceptible to bad weather since trophic resources are depressed at a time when both adults and young must be fed (Ojanen 1979, Wingfield et al. 1983). When food supply is reduced during extended periods of inclement weather, it is difficult for adults to feed themselves and the young, and death of the nestlings by starvation often results (Newton 1973). In other cases, the nest and eggs may be covered by transient snowstorms, flooded, or destroyed by hail without the adults being energetically or thermally stressed (Ehrlich et al. 1972, Kale 1973, Jehl and Hussell 1966, Myres 1955, Schroeder 1972, Morton 1976, Elkins 1983). Newton (1973) states that for some finches breeding in temperate zones, up to 80% of first clutches are lost to predators or severe storms. As spring progresses, the vegetation grows and thickens, thus providing more concealment for nests and shelter from storms. As a result, later nests tend to be more successful, although individuals still have a selective advantage if they can successfully raise young from early nests (Perrins 1970).

The catastrophic storms described above are not infrequently dismissed as "abnormal weather," giving the impression that they are highly unusual. However, examination of meteorologic data for at least north temperate regions indicates that "abnormal" storms occur about once every eight to ten years (Elkins 1983), and inclement weather of lesser severity, but nonetheless destructive, occurs on average every two years. Thus it appears likely that there is strong selection for individuals that respond to inclement weather, acclimate, at least briefly, and then readjust their breeding schedules after the storm has passed. In this chapter, the responses of birds to bad weather and other disruptive influences during the nesting phase (i.e., *after* gonadal development and onset of breeding), and the mechanisms by which individuals adjust their breeding schedule accordingly, will be discussed.

Environmental events that disrupt the nesting phase are called modifying factors (Wingfield 1980, 1983). Note that after disruption of the nesting phase, individuals often begin a renest attempt, i.e., they revert to earlier stages of the nesting phase and begin anew. This is distinct from multiple brooding in which subsequent nesting phases are initiated only after *successful* completion of the preceding nesting attempt. It is important to note this difference since the endocrine mechanisms underlying such marked transitions in behavior appear to be different. Disruption of the nesting phase, followed by subsequent renest attempts, must also be differentiated from the effects of supplementary factors that influence the preparatory phase of the reproductive cycle. In the latter, effects of deleterious environmental factors such as inclement weather act as inhibitors but do not require that an individual revert to earlier stages. Rather, inhibitors may result in a pause in or slowing down of the progression of the preparatory phase, but then the normal sequence of events continues as soon as conditions ameliorate (Wingfield 1984a). This is in contrast to the effects of modifying factors that disrupt the normal temporal progression of the nesting phase and require that individuals begin the nesting phase once again.

There are two main types of modifying factors, *direct* and *indirect*. Indirect modifying factors include predators that take nest and young or short severe storms that flood the nest or destroy it (e.g., hailstorms). Additionally, human disturbance frequently results in loss of the nest. Note that adults are not affected directly by these factors, except perhaps in terms of defense of the nest and young from predators. Loss of a mate to a predator, disease, or inclement weather is another example of an indirect modifying factor that requires radical readjustment of the surviving individual's reproductive cycle as a new mate is sought or the individual attempts to raise the young alone.

Direct modifying factors, on the other hand, include the effects of bad weather that result in a decrease of ambient temperature and/or reduction in available food. In these cases the adults often lose weight as they attempt to feed themselves and young. Further, since ambient temperature may also decline, it is often necessary to brood the young for longer periods, thus reducing time spent foraging for decreasing amounts of food. If adverse conditions persist, the adults may reach a point where energy reserves are reduced. In this case, the nest and young are abandoned before severe debilitation of the adults occurs. Thus the breeding attempt is aborted in favor of survival of the adults so that renesting can commence soon after the storm has passed.

II. Endocrine Adjustments During Renesting After Loss of Clutch or Brood

A. Temporal Patterns of Plasma Luteinizing Hormone (LH) and Sex Steroid Hormone Levels During Renesting

It is common for individuals that have lost the nest and young (as a result of both direct and indirect modifying factors) to renest following amelioration of environmental conditions. Such a strategy increases reproductive success, especially in those species that breed at higher latitudes or altitudes where the short summers allow successful raising of one brood only. In these instances, if the nest is lost sufficiently early in the season, then renesting follows, and individuals thus avoid total reproductive failure for that year. Since loss of the nest disrupts the normal temporal progression of the reproductive cycle, an extensive reorganization of reproductive function is required before renesting can begin. The endocrine changes associated with renesting have been investigated in three avian species thus far.

In the white-crowned sparrow, *Zonotrichia leucophrys,* there was a marked resurgence of luteinizing hormone (LH), testosterone, and testis mass after loss of the nest (Wingfield and Farner 1979; Figure 7-1A,B) and in females there was a coincident resurgence of LH and estrogens that culminated in production of a replacement clutch (Wingfield and Farner 1979; Figure 7-2). As soon as the clutch was complete, and incubation began, plasma levels of LH and sex steroid hormones declined rapidly in a manner similar to the transition from the sexual to parental subphases during the first brood (Wingfield and Farner 1978a,b, 1979). Essentially similar results have been obtained in female mallards, *Anas platyrhynchos,* after experimental removal of the eggs. Plasma levels of LH increased significantly within 12 hours of nest destruction (Donham et al.

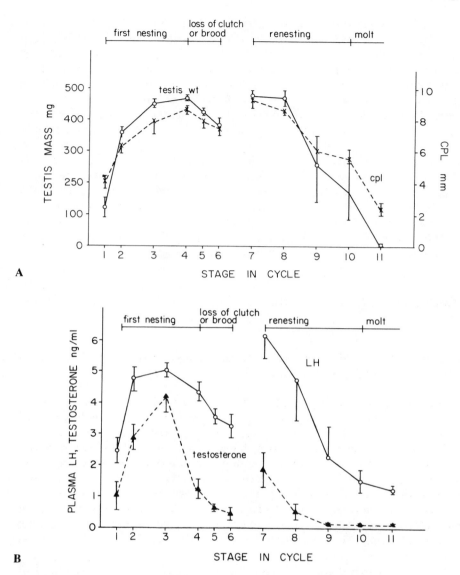

Figure 7-1. (A) Changes in testis mass and length of the cloacal protuberance (CPL) in renesting male white-crowned sparrows. Stages of the cycle are as follows: 1 = migration, 2 = arrival on territory, 3 = egg-laying stage for first clutch, 4 and 5 = incubation, 6 = feeding of young, 7 = egg-laying stage for replacement clutch after loss of previous nest, 8 = incubation, 9 = feeding young, 10 = feeding fledglings, 11 = molt. (B) Changes in plasma levels of luteinizing hormome (LH) and testosterone during renesting in male white-crowned sparrows. Stages in cycle as for (a). (From Wingfield and Farner 1979. Courtesy of Academic Press.)

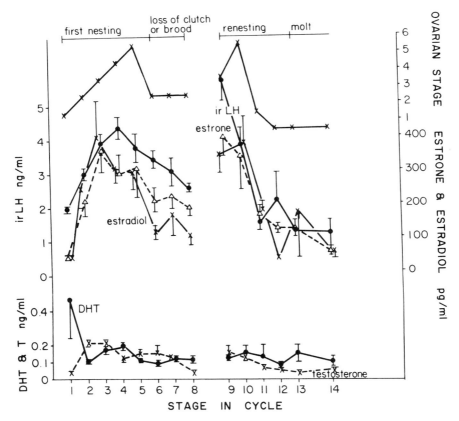

Figure 7-2. Changes in ovarian stage and plasma levels of luteinizing hormone (LH), estrone, estradiol, dihydrotestosterone (DHT), and testosterone (T) in renesting female white-crowned sparrows. Stages in cycle are as follows: 1 = migration, 2 = arrival on territory and pairing, 3 = beginning of yolk deposition in ovarian follicles, 4 = about to ovulate, 5 = egg-laying stage, 6 and 7 = incubation, 8 = feeding of young, 9 = yolk deposition for replacement clutch after loss of first, 10 = egg-laying stage for replacement clutch, 11 = incubation, 12 = feeding of young, 13 = feeding fledglings, 14 = molt. (From Wingfield and Farner, 1979. Courtesy of Academic Press.)

1976). Similarly, in song sparrows, *Melospiza melodia,* that lost nests to extensive flooding in 1982, there was a resurgence in plasma levels of LH and sex steroid hormones in both males and females as the replacement clutch was produced. As in male white-crowned sparrows, there was significant rise in plasma levels of testosterone in renesting male song sparrows (Wingfield 1985a) similar to that seen during the egg-laying period for the first clutch. This was unlike multiple brooding in which there was no second rise in testosterone during ovulation and oviposition for the second clutch after *successful* raising of the first brood (Wingfield 1984b). Note, however, that the changes in circulating levels of reproductive hormones in renesting females were essentially identical to the changes that occur during initiation of the first clutch (Wingfield 1985b, Wingfield and Farner 1979).

B. Renesting and Multiple Brooding

The environmental cues that regulate secretion of sex steroids in males during the egg-laying stage of the second brood (after successful completion of the first) compared with renesting remain unknown, at least in free-living species. The high levels of testosterone measured in the blood of males during the first egg-laying period appeared to be stimulated by the sexual behavior of the female (e.g., Moore 1982), and the decline of circulating testosterone as incubation begins may have been a result of the change in behavior of the female from sexual to parental (O'Connell et al. 1982). However, sexual behavior of the female failed to elicit an increase in plasma levels of testosterone in males during the egg-laying stage of the second brood but apparently did increase testosterone in males during renesting for a replacement clutch. Thus it appears that the presence of young somehow suppresses the response of the male to sexual behavior of the female.

It has been suggested that high levels of prolactin associated with the parental subphase suppress testosterone secretion during the egg-laying stage of the second brood (Wingfield and Farner 1979, Moore 1982; see also Camper and Burke 1977a,b). However, if the nest and young are lost to a predator or storm, then the stimulus for prolactin secretion (from the eggs or young) is lost; plasma levels decline, allowing a second rise in circulating levels of LH and sex steroid hormones (Wingfield et al. 1983, Wingfield 1985a,b).

This hypothesis has been tested in two domesticated species. In the canary, *Serinus canarius,* removal of the nest resulted in a decline of plasma levels of prolactin followed by a subsequent increase after the replacement clutch was produced (Goldsmith 1983). Similarly, in the domestic turkey, *Melagris gallopavo,* removal of the nest depressed prolactin levels in females. Replacement of the nest resulted in a rise of circulating prolactin (El Halawani et al. 1984). Although plasma levels of sex steroids were not measured, the data do support the hypothesis that loss of the nest and contents depress prolactin levels which could then allow a resurgence of circulating gonadal hormones. Anti gonadal actions of prolactin in birds have been suggested by investigations on Japanese quail, *Coturnix coturnix japonica,* and turkeys in which injections of prolactin into egg-laying females depressed LH and disrupted ovulatory cycles (Camper and Burke 1977a,b). These data lend further support to the hypothesis.

In another test, pairs of free-living song sparrows were sampled throughout a breeding season, and the nests of first broods removed to simulate loss to a predator or storm. Further samples were then collected after loss of the nest and during renesting. The prediction was that plasma levels of prolactin would decline after nest destruction, thus allowing a rise in sex steroid hormones. However, there were no changes in prolactin levels in males or females following loss of the nest (Wingfield and Goldsmith 1987). Plasma levels of testosterone in males and estradiol in females increased despite continued high levels of prolactin. Thus the hypothesis does not appear to be generally applicable.

C. Why Is There a Surge of Testosterone Secretion During Renesting but Not Multiple-Brooding?

The functional significance of the elevation of testosterone titer during renesting but not during multiple brooding is unclear, but one possible suggestion involves survival of young. High levels of testosterone, accompanied by increased territorial and "mate-guarding" aggression, are incompatible with parental behavior (Silverin 1980, Hegner and Wingfield 1987a) and would result in reduced care of the young, especially when males are feeding fledglings. Presumably, fitness of the male is enhanced if he invests more time feeding offspring, thus ensuring independence, rather than maximizing paternity of later clutches. Further, the chances of survival of young from first broods are greater than of young from subsequent broods (e.g., Perrins 1970). Fledglings of the first brood are a genetic investment in hand, whereas the second brood has to be incubated and fed. Thus the male is presented with an "evolutionary" dilemma of whether to feed the fledglings of the first brood until they are independent or invest more time "mate-guarding" the female as she becomes sexually receptive during the egg-laying stage of the second brood. Apparently selection has favored the former strategy, even though a male may be cuckolded by neighbors since he cannot "mate-guard" as efficiently as he did during ovulation and oviposition for the first brood. On the other hand, a parental male has several fledglings that are near to independence. Further, once these fledglings are independent and his mate is incubating the second clutch, he is able to seek extra-pair copulations and cuckold other males. This point is speculative but could be tested.

If the nest and eggs were lost or abandoned, then it would be of no disadvantage to the male if plasma levels of testosterone rose during the second egg-laying period, thus increasing levels of aggression and "mate-guarding" to ensure paternity of the replacement clutch (Wingfield 1985a). In those species that nest at high latitudes, the replacement clutch may be the only opportunity for reproductive success that season. Clearly, increased levels of testosterone and "mate-guarding" behavior would be an advantage during renesting, but a distinct disadvantage during multiple-brooding.

D. Hormonal Adjustments During Loss of Nest

What hormonal readjustments take place during the disruptive period prior to renesting, and by what mechanisms?

First, one must assume that before renesting can begin, an individual must be in a physiological state compatible with breeding. In Figure 7-3a it can be seen that body mass, fat depot, and plasma levels of corticosterone, often used an indicator of stress (e.g., Siegel 1980, Harvey et al. 1984), were similar in renesting males compared with those just prior to loss of the first nest. A similar trend was seen in female white-crowned sparrows (Figure 7-3b). In the latter case, body mass increased rapidly as new eggs were formed for the renest attempt (Wingfield and Farner 1979). Clearly, individuals had recovered body condition at least to levels prior to disruption of breeding. But what happens during the disruptive period?

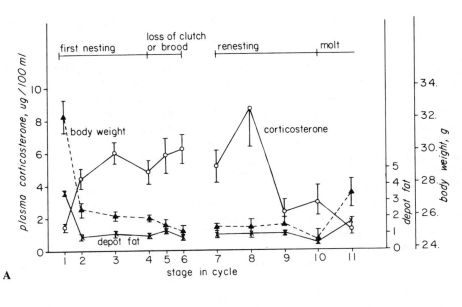

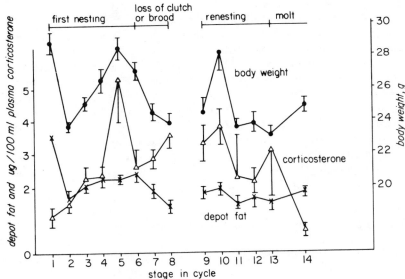

Figure 7-3. Changes in body weight, depot fat, and plasma levels of corticosterone in renesting male (A) and female (B) white-crowned sparrows. Stages in cycle in (A) and (B) as for Figures 7-1A and 7-2, respectively. (From Wingfield and Farner 1979. Courtesy of Academic Press.)

Since indirect modifying factors destroy the nest and young apparently without stressing the adult, then renesting can commence immediately. The endocrine responses to other indirect modifying factors, such as loss of a mate, are not known for any truly wild avian species, but it seems logical to assume that renesting will be held in abeyance until a new pair bond is formed. If the widowed individual attempts to raise young without the help of a mate, then it is possible that there could be profound changes in circulating levels of hormones associated with reproduction. In the pied flycatcher, *Ficedula hypoleuca,* some males obtained a second female during the breeding season, but the male assisted only his first mate in feeding the young. Silverin and Wingfield (1982) detected no differences in plasma levels of hormones between females feeding young with the help of a mate and females feeding young alone (analagous to a widowed female). However, a more in-depth analysis by Silverin (1985, 1983) indicated that female pied flycatchers raising young alone weighed less and had higher plasma levels of corticosterone than females feeding small broods with the assistance of a male. These data suggest that females show some signs of stress when feeding a brood alone. However, the plasma levels of corticosterone in single female pied flycatchers were not as high as those in truly stressed individuals, suggesting that feeding young alone is not severely debilitating. Indeed, it is quite clear that single females can raise young alone, at least to fledging.

Silverin (1985, 1983) also found a positive correlation of plasma levels of corticosterone and brood size (Figure 7-4). Female pied flycatchers feeding young in nests with two to three young had lower circulating levels of corticosterone than females feeding young in experimentally enlarged broods of eight or more. However, the levels of corticosterone in females feeding young in enlarged broods and the seasonal changes of corticosterone in blood were much less than those induced by stress. In more recent experiments in which brood size was manipulated in nest boxes of house sparrows, *Passer domesticus,* there were no correlations between plasma level of corticosterone in males and females and experimentally altered brood size (Hegner and Wingfield 1987b). However, the interbrood interval between fledging of the young and egg-laying for the next clutch was found to be significantly longer in females feeding enlarged broods. Although these data appear to contradict those of Silverin (1985, 1983), it is possible that these differences represent alternative reproductive strategies. It should also be noted that the house sparrow is a commensal of man, living in barnyards and farms where artificial and natural food is always abundant. Thus, it is possible that adult house sparrows do not have to work as hard to find food for an enlarged brood as the pied flycatcher which, as the name suggests, captures its food in the air or from foliage in trees.

The effects of direct modifying factors influence adults rather than the nest and young. Here, then, we might expect changes in endocrine function as individuals adjust to severe climatic conditions that may eventually result in abandonment of the nest. A common correlate of direct modifying factors, such as inclement weather, is reduced temperature and/or a decline in food resources. Low temperature and shortage of food are usually regarded as stressor stimuli (Siegel 1980, Harvey et al. 1984) and frequently accompany persistent inclement

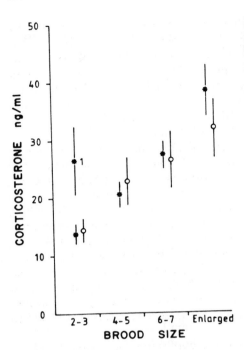

Figure 7-4. Plasma levels of corticosterone in relation to brood size in the female pied flycatcher. 1 = Females feeding young without assistance of a mate. (From Silverin [1982] with permission and courtesy from Academic Press and author.)

weather. Individuals with young may be particularly stressed in contrast to those subjected to bad weather before the nesting phase is underway (see Wingfield 1984a). Thus, the stress response could be responsible for abandonment of the nest under severe and persistent conditions.

Before going on to describe the stress concept, it is important to first of all point out that the very word itself is nonspecific and often confusing. In the literature it is not unusual to read phrases such as the "stress of reproduction." Such statements are rather misleading because Selye's (1956) definition of a stressor is an obnoxious stimulus which an individual must avoid, acclimate to, or move away from. If the individual can do none of these things, then its fitness will be greatly impaired. Given this definition, it is clearly inappropriate to consider reproduction an "obnoxious stimulus" and therefore a stress. The reproductive process is a normal function, energetically expensive, perhaps, but not stressful. Similar conclusions have been reached by King and Murphy (1985) concerning the energetic constraints of avian reproductive cycles. Part of the confusion lies in the fact that plasma levels of corticosterone normally tend to rise during the nesting phase (Wingfield and Farner 1978a,b, Wingfield 1984b), indicating, but not proving, an increase in stress.

III. The Concept of Stress

The responses of vertebrate organisms to stressor stimuli are well known, and birds fit the general pattern in showing a sustained increase in secretion of corticosterone in response to a variety of stressor stimuli such as high and low temperature, handling and minor surgical procedures, capture, anesthesia, inanition, etc. (e.g., Siegel 1980, Holmes and Phillips 1976, Etches et al. 1984, Harvey et al. 1984, Greenberg and Wingfield 1986). Increases in glucocorticosteroids are thought to mobilize energy reserves, especially protein, for combating stress during acclimation. Another common effect of stress is the decline of reproductive hormones such as LH, testosterone, and estradiol (see Assenmacher 1973, Wilson et al. 1979, Etches et al. 1984, Greenberg and Wingfield 1986).

Plasma levels of corticosterone increased during capture and handling stress in white-crowned sparrows, especially during the winter months (Wingfield et al. 1982). This was consistent with the effects of a variety of stressors on several domestic avian species (see above). However, changes in circulating levels of corticosterone in white-crowned sparrows were conspicuously lower than those of domesticated species. In some cases, e.g., in the breeding season, stress induced at most a sluggish increase in levels of corticosterone in contrast with conspicuous increases in winter (Wingfield et al. 1982). Similarly, pen-raised and feral California quail, *Lophortyx californicus,* responded only slightly (measured as depletion of ascorbic acid content in the adrenocortical tissue, a measure of steroid content) to cold, caging, and treatment with cortisol and adrenocorticotropic hormone (ACTH), which elicited marked responses in domestic species (Flickinger 1959). However, plasma levels of corticosterone in very young chickens (Newcomer 1959, Freeman and Manning 1979, Freeman and Flack 1980) and also in laying hens (Etches 1976) did not increase following handling and restraint, suggesting that even among domestic avian species there is variation in adrenal responses to stress. In contrast, other feral species such as the house sparrow and European starling, *Sturnus vulgaris,* did show very dramatic increases in plasma corticosterone levels in response to handling stress during the breeding season (see Dawson and Howe 1983, Hegner and Wingfield 1986).

The apparent reduced adrenal response to acute stress in some species, or individuals, could indicate either that they were maximally stressed at the onset of the experiment or that they responded rapidly to handling before a blood sample could be withdrawn. This is unlikely, however, for the white-crowned sparrow since plasma levels of corticosterone did not increase within two minutes of handling stress (Wingfield et al. 1982).

Seasonal changes, as of yet of unexplained origin, in responses of white-crowned sparrows to stress are noteworthy. In *Z.l. gambelli,* adrenocortical tissue regressed during the breeding season (Lorenzen and Farner 1964), which is consistent with the observation that the adrenal response to stress diminishes during the nesting phase. Possibly for *Z.l. gambelii,* which breeds at high latitudes and altitudes with brief summers, reduced adrenal response to acute environmental stress in summer may be adaptive in permitting the normal progression of reproductive function despite potentially severe climatic conditions. Environmental conditions in spring ameliorate rapidly and episodes of inclement weather are brief. Since maximal fitness depends upon the reproductive effort beginning as early as possible to avoid overlap with the first autumnal storms then reduced responsiveness to acute stressors in certain species in spring could be adaptive. More experimentation is required to clarify this issue.

IV. Environmental Cues as Direct Modifying Factors

Although some avian species apparently have evolved a "resistance" to *acute* stressor stimuli during the nesting phase, one can predict that a more *chronic* stimulus, such as a severe storm that persists for several days, would result in a marked endocrine stress response.

A. Temperature

Prolonged and severe storms may stress individuals during the nesting phase by depressing ambient temperature and trophic resources. In Figure 7-5 it can be seen that a storm in May 1980 at a study site in western Washington State resulted in a decline of ambient temperature for about eight days. In some instances, low environmental temperature is known to delay photoperiodically induced gonadal growth (Marshall 1949, Engels and Jenner 1956, Lofts and Murton 1966, Storey and Nicholls 1982) and in domesticated species resulted in a decline in reproductive function and an increase in adrenocortical activity (e.g., Assenmacher 1973, Edens and Siegel 1975, Huston 1975, Nir et al. 1975, Etches 1976), despite the fact that in all of these experiments the subjects had free access to food. On the other hand, in the white-crowned sparrow, fed *ad libitum*, low environmental temperature had only a marginal effect on photoperiodically induced gonadal growth (Lewis and Farner 1973). Similarly, in the

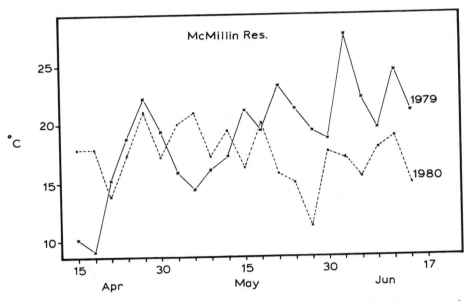

Figure 7-5. Changes in ambient temperature (3-day means) at McMillin Reservoir, western Washington State (42° N) during spring of 1979 and 1980. (From Wingfield et al. (1983), courtesy of American Ornithologist's Union.)

great tit, *Parus major*, low environmental temperature failed to influence gonadal growth induced by exposure to artificial long days (Suomalainen 1937). Even in sexually mature white-crowned sparrows, low temperature had no effect (Wingfield et al. 1982): A group of males exposed to low temperature (5°C) had the same plasma levels of LH, testosterone, and corticosterone as controls (23°C), suggesting that reduced ambient temperature was not stressful. This is unlike the responses of domesticated, and some feral, species and suggests that some birds are able to withstand the effects of low temperature if they have free access to food (Wingfield 1984a, Wingfield et al. 1982).

B. Food Availability

It is often assumed that reduced environmental temperature during storms is accompanied by a reduction of available food. This is particularly true for those species that feed themselves and their young primarily on invertebrates, especially insects. Many sparrows, including white-crowned and song sparrows, also feed their young on insects. Generally the reduction in food was extremely rapid and can be prolonged (Figure 7-6). After conditions ameliorate, food resources quickly returned to high levels. Thus, it is possible that decreased ambient temperature and heavy precipitation accompanying storms are stressful by reducing food resources. Impaired food intake, in turn, results in metabolic stress that is known to elevate plasma levels of corticosterone and depress circulating titers of gonadal hormones in domesticated species (see Siegel 1980, Nir et al. 1975, Jeronen et al. 1976, Beuving and Vonder 1978, Scanes et al. 1980, Assenmacher 1973, Assenmacher et al. 1965, Wilson et al. 1979). However, very few truly feral species have been studied in this regard.

Male white-crowned sparrows subjected to a restricted diet showed attenuated cycles in plasma levels of LH, basal levels of testosterone, and elevated circulating levels of corticosterone compared with controls (Figure 7-7). Note that in this and the next experiment, all birds were maintained at an ambient temperature of 23°C. In another experiment, mature male white-crowned sparrows had food removed from their cages for one day and were sampled after 24 hours of food deprivation. This paradigm simulates the effects of a severe storm on food resources without influencing ambient temperature. Males deprived of food for 24 hours had basal levels of testosterone and increased titers of corticosterone. Curiously, however, plasma levels of LH were not affected (Figure 7-8). One week after food was returned, plasma levels of testosterone and corticosterone were similar to those of controls. Thus it seems to be quite clear that under laboratory experimentation, inanition results in marked increases of corticosterone and a reduction of testosterone levels. Effects on LH appear to be more equivocal.

C. Effects of Corticosteroids on Reproductive Function

It has been suggested that stress-induced elevations of ACTH and corticosteroids inhibit gonadal function directly. This hypothesis has attracted consid-

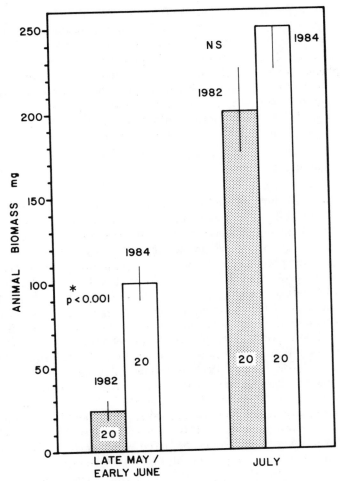

Figure 7-6. Comparison of animal biomass (i.e., invertebrate food available to breeding song sparrows assessed by sweep net collections in the breeding habitat) in spring and summer of 1982 and 1984. Food data were not available for 1981, but spring and summer of 1984 were very similar to 1981 and there were no severe storms.

erable attention in mammals and birds (Selye 1956, Christian et al. 1965, Cain and Lien 1985, Assenmacher 1973, Greenberg and Wingfield 1986). However, such a simplistic view is not supported fully by the literature. On the positive side, ACTH and corticosteroids have been shown to cause morphological changes suggestive of hyperadrenocortical function and atrophy of the reproductive organs of cocks and hens (Flickinger 1966a,b). More recently, Wilson and Follett (1975) in tree sparrows, *Spizella arborea*, and Deviche et al. (1979) in domestic ducks, have shown that administration of corticosterone, whether systemically or directly into the basal hypothalamus, depressed plasma levels

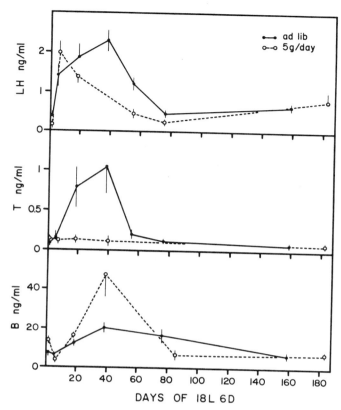

Figure 7-7. Effects of chronic food restriction on plasma levels of luteinizing hormone (LH), testosterone (T), and corticosterone (B) in photostimulated male white-crowned sparrows. (From Wingfield and Farner 1980a.)

of LH. In the bobwhite quail, *Colinus virginianus*, Cain and Lien (1985) have accumulated considerable evidence that corticosterone treatment resulted in decreased testis mass and sperm production and reduced the number of eggs laid, as well as decreased ovary and oviduct mass. Field observations of the western gull, *Larus occidentalis*, support the above hypothesis since plasma levels of corticosterone were highest in females that had low body mass just prior to the breeding season in a year in which available food was only 10% of normal (Wingfield 1983). These females failed to breed and plasma levels of LH and estrogen remained basal.

On the negative side, in the domestic duck, Deviche et al. (1980) were unable to depress plasma levels of LH and FSH following injections of ACTH, despite the fact that such treatment resulted in significant elevation of circulating levels of corticosterone. Similarly, stressor stimuli, other than inanition, that increase plasma levels of corticosterone in male white-crowned sparrows do not necessarily result in the inhibition of LH and androgen secretion and reproductive

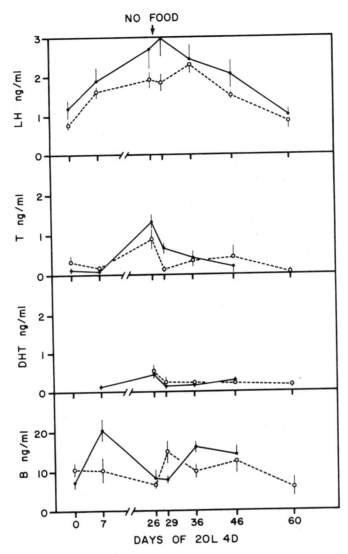

Figure 7-8. Effects of acute food restriction on plasma levels of luteinizing hormone (LH), testosterone (T), dihydrotestosterone (DHT), and corticosterone (B) in photo-stimulated male white-crowned sparrows. (From Wingfield and Farner 1980a.)

function (Wingfield et al. 1982). This is in contrast to the effects of food deprivation in which a dramatic decline of testosterone is observed (Figure 7-7 and 7-8). Furthermore, injections of ACTH and implants of corticosterone failed to influence photoperiodically induced rises in LH and testosterone in male white-crowned sparrows (Wingfield and Farner 1980a).

Despite these contradictions in the literature, it is nevertheless possible to hypothesize that the stressful effects of storms during the nesting phase result in an elevation of circulating corticosterone to mobilize energy reserves. In addition, increased adrenocortical activity suppresses reproductive function by depressing the secretion of sex steroid hormones, and possibly also LH, thus resulting in a disruption of the nesting phase. After conditions ameliorate, corticosterone levels abate, thus releasing an inhibition on sex hormone secretion, and renesting ensues. The next question is, do these mechanisms operate in free-living individuals in response to a natural storm?

V. Endocrine Responses to Inclement Weather During the Nesting Phase of the Reproductive Cycle

A. Effects of Weather on Free-Living Populations

In recent years, field-endocrine techniques have allowed opportunistic investigations of the effects of unpredictable storms on reproductive function in free-living populations of birds. One such storm occurred in 1980 during the breeding season of a population of white-crowned sparrows, Z. l. pugetensis, in western Washington State compared with a relatively storm-free year of 1979. The weather in early spring in both 1979 and 1980 was warm and sunny with temperatures usually remaining above 15°C (see Figure 7-5). In both years, white-crowned sparrows were feeding nestlings or fledglings by late May. At this time in 1980, however, there was a precipitous decline in temperature, to as low as 10°C below that recorded at the same time in 1979 (Figure 7-5), accompanied by heavy overcast and precipitation. In June, weather conditions ameliorated, and although temperatures remained slightly below those for same time in 1979, they were well within the normal range for this time of year.

In birds sampled during late May 1980, plasma levels of corticosterone were elevated above those of birds sampled at the same time in 1979 and at the same stage of the nesting phase (Figure 7-9). By July 1980, when birds were feeding nestlings of a renest attempt, corticosterone levels had declined. Subcutaneous fat deposits were also virtually depleted in birds sampled during the storm of 1980 but were restored by July (Figure 7-9). Thus it appears that male white-crowned sparrows showed a typical stress response during the storm in 1980. Curiously, however, plasma levels of LH and testosterone were unaffected (Figure 7-9), perhaps because these levels had already declined from high vernal levels as is typical of monogamous species in which males show parental behavior (see Wingfield 1980, 1983, 1984b). Furthermore, it seems likely that nests were abandoned as no fledglings were observed until after a renest attempt in June and July (Wingfield et al. 1983).

Remarkably similar responses to a storm during the nesting phase have been presented by Wingfield (1985a,b) for male song sparrows breeding in the mid-Hudson Valley of New York State. In May and June 1982 there were periods of intense precipitation (up to 50 mm per day) preceded by a 5 to 10°C drop in

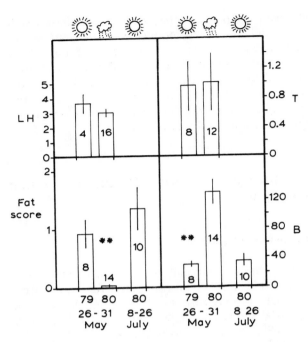

Figure 7-9. Effects of inclement weather on plasma levels of luteinizing hormone (LH), testosterone (T), corticosterone (B) and fat score in free-living male white-crowned sparrows. Stylized sun represents fair weather; rain cloud indicates periods of inclement weather. (From Wingfield et al. (1983), courtesy of American Ornithologist's Union.)

temperature. The same period in 1981 was relatively dry and warm with no severe storms. In early June 1982, weather conditions ameliorated and were similar to conditions at the same time in 1981. As was found with white-crowned sparrows, plasma levels of LH and testosterone in male song sparrows did not decline during the period of inclement weather in 1982 (Wingfield 1985a). Plasma levels of testosterone increased markedly during the egg-laying stage of the renesting attempt (see also the section on renesting above). There was no observable depletion of fat depot in male song sparrows during the storms of 1982, possibly because fat deposits normally are virtually zero at this time of year. However, there was a significant decline in body mass and a marked elevation of circulating corticosterone levels in males during the stormy period of 1982 compared with the same period in 1981. Furthermore, these males abandoned their nests even though virtually all birds sampled were feeding young and fledglings. Some males were observed up to several hundered meters away from their breeding territories, suggesting at least partial breakdown of territorial behavior. However, as soon as the storms abated, males resumed territorial and courtship behavior behavior and renest attempts began (Wingfield 1985a).

B. Sex Differences

The endocrine responses of female song sparrows to the stormy period in late May and early June differed from those of males (Wingfield, 1985b). In males, body mass declined and corticosterone levels rose during the period of inclement weather, whereas in females there were no differences between the two years, suggesting that they were not stressed by storms. In fact, plasma levels of corticosterone tended to be higher in the year of fair weather than in the stormy period. There was, however, a significant decline in plasma levels of estradiol during the stormy period and a resurgence as females prepared to lay a second clutch. This second increase in estradiol, and also LH, is typical of the temporal pattern of multiple-brooded species (see Wingfield and Farner 1980b, Wingfield 1983) and of renesting attempts after loss of a clutch or brood (Wingfield and Farner 1979; see also above).

The mechanisms by which estradiol levels are depressed during the May and June 1982 storms are unknown. A stress response is unlikely because there were no differences in body mass, fat depot, and corticosterone levels between the two years. Curiously, however, males did appear to be stressed at this time. This sex difference in response of song sparrows to the same storm may have a basis for explanation if one considers patterns of parental behavior in this species. When the storms occurred, virtually all pairs of song sparrows were either feeding large nestlings or fledglings (mostly the latter). After the young had fledged, females began preparations for a second brood and males fed the fledglings (Nice 1943). Thus, storms striking at a time when males have very low fat reserves (Wingfield 1985a), and have to feed themselves and young as well, could result in severe energetic stress. Females, on the other hand, are not feeding fledglings intensively and have much larger fat depots through much of the nesting phase (Wingfield 1985a,b). Thus, at this time in the nesting phase, females may be much less susceptible to stress.

Additional support for this hypothesis comes from Wingfield et al. (1982) who showed that in free-living populations of white-crowned sparrows, females are much more resistant to the effects of at least acute stress stimuli during the nesting phase of the reproductive cycle (see also the section on stress above). It would, however, be interesting to observe the effects of inclement weather on the endocrine response of females during incubation or brooding of newly hatched young and determine what advantages are conferred by having a large fat depot at this time. Since male sparrows apparently do not incubate and rarely brood young, one would predict that at this time a storm would result in stress of females but not males. Thus, the endocrine responses of free-living birds to inclement weather depend not only on the intensity of the storm but also on the sex, mating system, and relative roles of the male and female during the parental phase.

C. Conflict of Laboratory and Field Evidence

Clearly, the evidence suggests that severe storms during the nesting phase of the reproductive cycle in males, at least when individuals are incubating or feeding young, can be stressful in terms of loss of body condition and hyperactivity of the adrenocortical tissue. However, in contrast to the observations in laboratory investigations on the effects of metabolic stress on both domesticated and feral species (see above), plasma levels of LH and testosterone do not appear to be affected. It should be noted that laboratory experiments on the effects of stress on reproductive function are frequently severe and allow the subject no other recourse (i.e., food is restricted or ambient temperature decreased chronically). However, under natural conditions, individuals have alternative strategies that they can pursue if conditions are severe. Thus, when ambient temperatures and/or decreased food availability results in potentially stressful conditions while feeding young, abandonment of the nest and a temporary period of halted reproductive activity would tend to reduce the potential stress on an individual. In this way the effects of unfavorable environmental conditions are avoided *before* the bird becomes severely stressed and the reproductive system inhibited. The reproductive system is maintained in a near functional state, allowing rapid renesting attempts once conditions once again become favorable. Note that if reproductive function was inhibited (as indicated in laboratory studies outlined above) then gonadal involution would occur that would require several weeks to reverse when condtions ameliorated. Obviously, caution should be exercised when extrapolating laboratory data to individuals living under natural conditions.

How, then, do we rationalize the breakdown of reproductive behavior that may result not only in abandonment of the nest and young, but also the territory? Recent evidence suggests that corticosterone may influence behavior directly as well as regulate metabolism.

VI. Corticosterone and Behavior During Stressful Episodes

Over the past few years, several investigations have implicated a role for corticosterone in the regulation of submissive agonistic behaviors in mice (Roche and Leshner 1979, Nock and Leshner 1976), increased food intake (Leibowitz et al. 1984) in rats, and hyperphagia in castrated pheasants, *Phasianus colchicus,* (Nagra et al. 1963). It is possible that corticosterone may also influence reproductive behavior during stressful episodes. To test this, captive male white-crowned sparrows were treated with subcutaneous implants of corticosterone, an empty implant as control, or metyrapone (a drug that inhibits 11β-hydroxylase—an enzyme essential for the biosynthesis of corticosterone). Metyrapone reduces plasma levels of corticosterone to the lower physiological range for birds, and well below the normal level for mature male sparrows. Foraging behavior was then observed over a period of several days following treatment. Foraging behavior was described as time spent actually feeding at the food cup and time spent searching for food (stereotyped scratching and searching movements). Corticosterone tended to increase time spent foraging over preimplant levels whereas there was no change in controls. Conversely, metyrapone-treated birds showed a decline in foraging to levels significantly lower than in corticosterone-treated birds (Gearhart et al. 1987). Thus it appears that there is a positive correlation between plasma levels of corticosterone and time spent foraging. However, metyrapone could depress feeding by a pharmacological action not related to a decrease in corticosterone secretion. To test this, a group of metyrapone-treated birds were given corticosterone, which resulted in an increase in foraging behavior to levels identical to controls (Gearhart et al. 1987). These data suggest that metyrapone does indeed decrease foraging behavior by impairing corticosterone secretion rather than by a pharmacological action.

The effects of storms include a depression of trophic resources. Do plasma levels of corticosterone influence foraging behavior during restriction of food intake? In a separate experiment, food was restricted for one day and foraging behavior observed in birds given corticosterone, metyrapone, or control implants. Immediately after implant, the distributions of time spent foraging were as described above, with corticosterone-treated birds showing most foraging and metyrapone-treated birds showing the least. After one day of food restriction, levels of foraging went up in all groups, which is hardly surprising since the birds were hungry. However, the increase in metyrapone-treated birds was less than in controls and corticosterone-treated birds (Gearhart et al. 1987). One week after food was returned, foraging levels had declined, but the same relationship with treatment holds. Clearly, then, corticosterone is important for increased foraging behavior during stressful episodes. Thus it is possible that stress-induced increases in corticosterone may redirect an individual's behavior away from reproduction toward foraging and survival, and without any concomitant changes in plasma levels of testosterone.

Although corticosterone appears to influence foraging behavior, the question

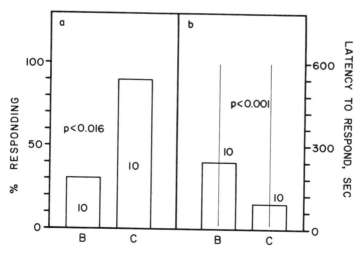

Figure 7-10. Effects of corticosterone (B) on responses of free-living and territorial male song sparrows to a simulated territorial intrusion. C = control. (From Wingfield and Silverin 1986. Courtesy of Academic Press.)

still remains as to whether territorial behavior is affected. To test this, free-living male song sparrows were given subcutaneous implants of corticosterone, and a second group was given empty implants as controls. Twenty-four hours after the implants were administered, the territorial behavior of each implanted bird was tested by placing a caged male song sparrow in the center of the subject's territory and playing back conspecific song to simulate a territorial intrusion (Wingfield and Silverin 1986). Territorial males usually respond vigorously and attempt to drive away the intruder. Thus the frequency and intensity of territorial response can be easily quantified over, for example, a 10-minute test period. Results are presented in the Figures 7-10–7-12. In Figure 7-10 it can be seen that only three out of ten corticosterone-treated birds responded to simulated territorial intrusion, whereas nine out of ten controls responded.

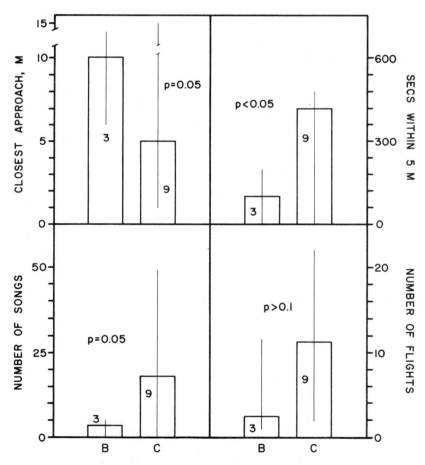

Figure 7-11. Effects of corticosterone (B) on territorial behavior of free-living male song sparrows. C = control. (From Wingfield and Silverin 1986. Courtesy of Academic Press.)

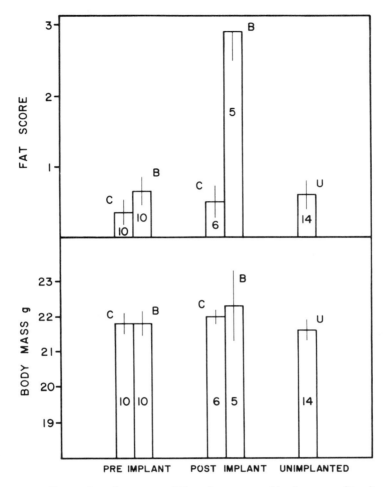

Figure 7-12. Effects of corticosterone (B) on fat score and body mass of territorial, free-living male song sparrows. C = control, U = untreated. (From Wingfield and Silverin 1986. Courtesy of Academic Press.)

The latency to respond in controls was also significantly lower than in corticosterone-treated birds. If the behavior of the three experimental birds that did respond was compared with that of controls, corticosterone-treated birds did not approach the intruder as closely as controls, spent less time within 5 meters, gave fewer songs, and tended to give fewer flight displays, suggesting that their territorial behavior had been subdued (Figure 7-11). Of the seven that did not respond, all were resighted within 100 meters of their territories, and some were recaptured. In Figure 7-12 it can be seen that corticosterone-treated males showed a huge increase in stored fat over controls and unimplanted males. This increase in fat is consistent with hyperphagia since, as noted above, corticosterone also appears to increase time spent foraging. Note that there is no increase in body weight in corticosterone-treated males, suggesting that despite the deposition of fat, protein is being mobilized (a well-known metabolic action of corticosterone).

The data presented above suggest strongly that chronic stressful episodes, such as severe storms, induce an increase in corticosterone that mobilizes energy reserves to combat reduced food intake. In addition, corticosterone redirects an individual's behavior away from reproduction and defense of a territory toward survival and increased foraging, without a concomitant decrease in plasma levels of LH and testosterone. This may be highly adaptive because if LH and testosterone secretions were inhibited, then gonadal regression would ensue. After weather conditions ameliorate, the time taken to reverse gonadal regression and produce a mature gonad for the renest attempt would dramatically decrease the time for successful completion of a subsequent brood. Since LH and testosterone are not affected, the gonad is maintained in a functional, or near functional, state so that renesting can begin *immediately* after weather conditions again permit the nesting phase to ensue.

Acknowledgments. Preparation of this manuscript was aided by grant number DCB 8316155 from the National Science Foundation, a Charles H. Revson Foundation Fellowship in Biomedical Research, and an Irma T. Hirschl Foundation Research Career Development Award to the author. I am also indebted to Christine Levesque who drafted the figures.

References

Assenmacher I (1973) The peripheral endocrine glands. In: Farner DS, King JR (eds) Avian Biology, Vol. 3. Academic, New York, pp. 183–286.

Assenmacher I, Tixier-Vidal A, Astier H (1965) Effet de la sous alimentation et du jeûne sur la gonadostimulation du canard. Ann Endocrinol 26:1–26.

Beuving B, Vonder GMA (1978) Effects of stressing factors on corticosterone levels in plasma of laying hens. Gen Comp Endocrinol 35:153–159.

Cain JR, Lien RJ (1985) A model for drought inhibition of bobwhite quail (*Colinus virginianus*) reproductive systems. Comp Biochem Physiol 82A:925–930.

Camper PM, Burke WH (1977a) The effects of prolactin on the gonadotropin induced rise in serum estradiol and progesterone in the laying turkey. Gen Comp Endocrinol 32:72–77.

Camper PM, Burke WH (1977b) The effect of prolactin on reproductive function in female Japanese quail (*Coturnix coturnix japonica*). Poultry Sci 56:1130–1134.

Christian JJ, Lloyd JA, Davis DE (1965) The role of endocrines in the self regulation of mammalian populations. Rec Prog Horm Res 21:501–571.

Cohen IR, Mann DR (1979) Seasonal changes associated with puberty in female rat. Effect of photoperiod and ACTH administration. Biol Reprod 20:757–762.

Dawson A, Howe PD (1983) Plasma corticosterone in wild starlings (*Sturnus vulgaris*) immediately following capture and in relation to body weight during the annual cycle. Gen Comp Endocrinol 51:303–308.

Deviche P, Balthazart J, Heyns W, Hendrick J-C (1979) Inhibition of LH plasma levels by corticosterone administration in the male duckling (*Anas platyrhynchos*). IRCS Med Sci 7:622.

Deviche P, Balthazart J, Heyns W, Hendrick J-C (1980) Endocrine effects of castration followed by androgen replacement and ACTH injections in the male domestic duck (*Anas platyrhynchos* L). Gen Comp Endocrinol 41:53–61.

Donham RS, Dane CW, Farner DS (1976) Plasma luteinizing hormone and the development of ovarian follicles after loss of clutch in female mallards (*Anas platyrhynchos*). Gen Comp Endocrinol 29:152–155.

Edens FW, Siegel HS (1975) Adrenal responses to high and low ACTH response lines of chickens during acute heat stress. Gen Comp Endocrinol 25:64–73.

Ehrlich PR, Breedlove DE, Brussard PF, Sharp MA (1972) Weather and regulation of subalpine populations. Ecology 53:243–247.

El Halawani ME, Burke WH, Millam JR, Fehrer SC, Hargis BM (1984) Regulation of prolactin and its role in gallinaceous bird reproduction. J Exp Zool 232:521–530.

Elkins N (1983) Weather and Bird Behaviour. Poyser, Calton.

Engels WL, Jenner CE (1956) The effect of temperature on testicular recrudescence in juncos at different photoperiods. Biol Bull 110:129–137.

Etches RJ (1976) A radioimmunoassay for corticosterone and its application to the measurement of stress in poultry. Steroids 28:763–773.

Etches RJ, Williams JB, Rzasa J (1984) Effects of corticosterone and dietary changes in the hen on ovarian function, plasma LH and steroids and the response to exogenous LH-RH. J Reprod Fertil 70:121–130.

Farner DS (1985) Annual rhythms. Ann Rev Physiol 47:65–82.

Farner DS, Gwinner E (1980) Photoperiodicity, circannual and reproductive cycles. In: Epple A, Stetson MH (eds) Avian Endocrinology. Academic, New York, pp. 331–366.

Flickinger DP (1959) Adrenal responses of California quail subjected to various physiologic stimuli. Proc Soc Exp Biol Med 100:23–25.

Flickinger GL (1966a) Effect of prolonged ACTH administration on gonads of sexually mature chickens. Poultry Sci 45:753–761.

Flickinger GL (1966b) Responses of testes to social interactions among grouped chickens. Gen Comp Endocrinol 6:89–98.

Freeman BM, Flack IH (1980) Effects of handling on plasma corticosterone concentrations in the immature domestic fowl. Comp Biochem Physiol 66A:77–81.

Freeman BM, Manning ACC (1979) Stressor effects of handling on immature fowl. Res Vet Sci 26:223–226.

Gearhart CE, Wingfield JC, Farner DS (1987) Effects of corticosterone on the foraging behavior of white-crowned sparrows, *Zonotrichia leucophrys gambelii* (submitted).

Gessamen JA, Worthen GL (1982) The effects of weather on avian mortality. Utah State University Printing Services, Logan.

Goldsmith AR (1983) Prolactin and avian reproductive cycles. In: Balthazart J, Pröve E, Gilles R (eds) Hormones and Behaviour in Higher Vertebrates. Springer-Verlag, Berlin, pp. 375–387.

Green GH, Greenwood JJD, Lloyd CS (1977) The influence of snow conditions on the date of breeding of wading birds in northeast Greenland. J Zool Lond Ser A 183:311–328.

Greenberg N, Wingfield JC (1987) Stress and reproduction: Reciprocal relationships. In: Norris DO, Jones RE (eds) Hormones and Reproduction in Fishes, Amphibians and Reptiles. Plenum, New York pp. 461–503.

Harvey S, Phillips JG, Rees A, Hall TR (1984) Stress and adrenal function. J Exp Zool 232:633–646.

Hegner RE, Wingfield JC (1986) Behavioral and endocrine correlates of multiple-brooding in the semicolonial house sparrow *Passer domesticus*. 1. Males. Horm Behav 20:294–312.

Hegner RE, Wingfield JC (1987a) Effects of brood size manipulation on parental investment, breeding success, and reproductive endocrinology of house sparrows. Auk (in press).

Hegner RE, Wingfield JC (1987b) Effects of experimental manipulation of testosterone levels on parental investment and breeding success in male house sparrows. Auk (in press).

Holmes WN, Phillips JG (1976) The adrenal cortex of birds. In: Chester-Jones I (ed) General, Comparative and Clinical Endocrinology of the Adrenal Cortex. Academic, New York, pp. 293–420.

Huston TM (1975) The effects of environmental temperature on fertility of the domestic fowl. Poultry Sci 54:1180–1183.

Jehl JR Jr, Hussell DJT (1966) Effects of weather on reproductive success of birds at Churchill, Manitoba. Arctic 19:185–191.

Jeronen E, Isometsa R, Hissa R, Pyornila A (1976) Effect of acute temperature stress on the plasma catecholamine, corticosterone, and metabolite levels in the pigeon. Comp Biochem Physiol 55C:17–22.

Kale HW (1973) House finch nest abandoned after snow. Wilson Bull 85:87.

King JR, Murphy ME (1985) Periods of nutritional stress in the annual cycles of endotherms: fact or fiction. Am Zool 25:955–964.

Lack D (1968) Ecological Adaptations for Breeding in Birds. Methuen, London.

Leibowtiz S, Roland CR, Hor L, Squillari V (1984) Noradrenergic elicited feeding via paraventricular nucleus is dependent upon circulating corticosterone. Physiol Behav 32:857–864.

Lewis RA, Farner DS (1973) Temperature modulation of photoperiodically induced vernal phenomena in white-crowned sparrows (*Zonotrichia leucophrys*). Condor 75:279–286.

Lofts B, Murton RK (1966) The role of weather, food and biological factors in timing the sexual cycle of woodpigeons. Brit Birds 59:261–280.

Lorenzen LC, Farner DS (1964) An annual cycle in the adrenal gland of the white-crowned sparrow, *Zonotrichia leucophrys gambelii*. Gen Comp Endocrinol 4:253–263.

Marshall AJ (1949) Weather factors and spermatogenesis in birds. Proc R Zool Soc 119:711–716.

Moore MC (1982) Hormonal response of free-living male white-crowned sparrows to experimental manipulation of female sexual behavior. Horm Behav 16:323–329.

Morton ML (1976) Adaptive strategies of *Zonotrichia* breeding at high altitude. In: Frith HJ, Calaby JH (eds) Proc 16th Int Ornithol Cong. Australian Acad Sci, Canberra, pp. 322–336.

Myres MT (1955) The breeding of blackbird, song thrush, and mistle thrush in Great Britain. Part 1. Breeding seasons. Bird Study 1:2–24.

Nagra B, Breitenbach RP, and Meyer RK (1963) Influence of hormones on food intake and lipid deposition in castrated pheasants. Poultry Sci 42:770–775.

Newcomer WS (1959) Adrenal and blood Δ_5-3-ketocorticosteroids following various treatments in the chick. Am J Physiol 196:276–278.

Newton I (1973) Finches. Taplinger, New York.

Nice MM (1937) Studies of the life history of the song sparrow. 1. A population study of the song sparrow and other passerines. Trans Linn Soc NY 4:1–239.

Nice MM (1943) Studies in the life history of the song sparrow. 2. The behavior of the song sparrow and other passerines. Trans Linn Soc NY 6:1–388.

Nir ID, Yam D, Perek M (1975) Effects of stress on the corticosterone content of the blood plasma and the adrenal gland content of intact and bursectomized *Gallus domesticus*. Poultry Sci 54:2101–2110.

Nock BL, Leshner AI (1976) Hormonal mediation of the effects of defeat on agonistic responding in mice. Physiol Behav 17:111–119.

O'Connell ME, Silver R, Feder HH, Reboulleau C (1981) Social interactions and androgen levels in birds. II. Social factors associated with a decline in plasma androgen levels in male ring doves (*Streptopelia risoria*). Gen Comp Endocrinol 44:464–469.

Ogle TF (1977) Modification of serum luteinizing hormone and prolactin concentration by corticotropin and adrenalectomy in ovariectomized rats. Endocrinology 101:494–497.

Ojanen M (1979) Effect of a cold spell on birds in northern Finland. Ornis Fenn 56:148–155.

Perrins CJ (1970) The timing of birds' breeding seasons. Ibis 112:242–255.

Pullainen E (1978) Influence of heavy snow fall in June 1977 on the life of birds in NE Finnish Lapland. Aquilo Ser Zool 18:1–14.

Riddle O, Honeywell HE (1924) Studies on the physiology of reproduction in birds. XVIII. Effects of the onset of cold weather on blood sugar and ovulation rate in pigeons. Am J Physiol 67:337–345.

Roche KE, Leshner AI (1979) ACTH and vasopressin treatments immediately after a defeat increase future submissiveness in male mice. Science 204:1343–1344.

Scanes CG, Merrill GF, Ford R, Mauser P, Horowitz C (1980) Effects of stress (hypoglycemia, endotoxin and ether) on the peripheral circulating concentration of corticosterone in the domestic fowl (*Gallus domesticus*). Comp Biochem Physiol 66C:183–186.

Schroeder, MH (1972) Vesper sparrow nests abandoned after snow. Wilson Bull 84:98–99.

Selye H (1956) The Stress of Life. McGraw-Hill, New York.

Siegel HS (1980) Physiological stress in birds. BioScience 30:529–534.

Silverin B (1980) Effects of long-acting testosterone treatment on free-living pied flycatchers, *Ficedula hypoleuca*, during the breeding period. Anim Behav 28:906–912.

Silverin B (1982) Endocrine correlates of brood size in adult Pied flycatchers, *Ficedula hypoleuca*. Gen Comp Endocrinol 47:18–23.

Silverin B (1983) Population endocrinology of the female pied flycatcher, *Ficedula hypoleuca*. In: Balthazart J, Pröve E, Gilles R (eds) Hormones and Behaviour in Higher Vertebrates. Springer-Verlag, Berlin, pp. 388–397.

Silverin B (1985) Cortical activity and breeding success in the pied flycatcher, *Ficedula hypoleuca*. In: Lofts B, Homes WN (eds) Current Trends in Comparative Endocrinology. Hong Kong University Press, Hong Kong, pp. 429–431.

Silverin B, Wingfield JC (1982) Patterns of breeding behavior and plasma levels of hormones in a free-living population of pied flycatchers, *Ficedula hypoleuca*. J Zool Lond Ser A 198:117–129.

Storey CR, Nicholls TJ (1982) Low environmental temperature delays photoperiodic induction of avian testicular maturation and the onset of postnuptial refractoriness. Ibis 124:172–174.

Suomalainen H (1937) The effect of temperature on the sexual activity of non-migratory birds stimulated by artificial lighting. Ornis Fenn 14:108–112.

Tompa FS (1971) Catastrophic mortality and its population consequences. Auk 88:753–759.

Whitmore RC, Mosher JA, Frost HH (1977) Spring migrant mortality during unseasonable weather. Auk 94:778–781.

Wilson EK, Rogler JC, Erb RE (1979) Effect of sexual experience, location, malnutrition, and repeated sampling on concentrations of testosterone in blood plasma of *Gallus domesticus* roosters. Poultry Sci 58:178–186.

Wilson FE, Follett BK (1975) Corticosterone-induced gonadosuppression in photostimulated tree sparrows. Life Sci 17:1451–1456.

Wingfield JC (1980) Fine temporal adjustment of reproductive functions. In: Epple A, Stetson MH (eds) Avian Endocrinology. Academic, New York, pp. 367–389.

Wingfield JC (1983) Environmental and endocrine control of reproduction: an ecological approach. In: Mikami S-I, Ishii I, Wada M (eds) Avian Endocrinology: Environmental and Ecological Aspects. Jap Sci Soc, Tokyo, Springer-Verlag, Berlin, pp. 265–288.

Wingfield JC (1984a) Influence of weather on reproduction. J Exp Zool 232:589–594.

Wingfield JC (1984b) Environmental and endocrine control of reproduction in the song sparrow, *Melospiza melodia*. 1. Temporal organization of the breeding cycle. Gen Comp Endocrinol 56:406–416.

Wingfield JC (1985a) Influences of weather on reproduction in male song sparrows, *Melospiza melodia*. J Zool Lond Ser A 205:525–544.

Wingfield JC (1985b) Influence of weather on reproduction in female song sparrows, *Melospiza melodia*. J Zool Lond Ser A 205:545–558.

Wingfield JC, Farner DS (1978a) The endocrinology of a naturally breeding population of the white-crowned sparrow, *Zonotrichia leucophrys pugetensis*. Physiol Zool 51:188–205.

Wingfield JC, Farner DS (1978b) The annual cycle of plasma irLH and steroid hormones in feral populations of the white-crowned sparrow, *Zonotrichia leucophrys gambelii*. Biol Reprod 19:1046–1056.

Wingfield JC, Farner DS (1979) Some endocrine correlates of renesting after loss of clutch or brood in the white-crowned sparrow, *Zonotrichia leucophrys gambelii*. Gen Comp Endocrinol 38:322–331.

Wingfield JC, Farner DS (1980a) Endocrinologic and reproductive states of bird populations under environmental stress. Report to the Environmental Protection Agency, Contract number cc 699095, 125 pp.

Wingfield JC, Farner DS (1980b) Control of seasonal reproduction in temperate zone birds. Prog Reprod Biol 5:62–101.

Wingfield JC, Goldsmith AR (1987) Changes in plasma levels of prolactin and sex steroid hormones in relation to multiple-brooding and renesting in free-living song sparrows, *Melospiza melodia*. submitted.

Wingfield JC, Silverin B (1986) Effects of corticosterone on territorial behavior of free-living male song sparrows *Melospiza melodia*. Horm Behav 20:405–417.

Wingfield JC, Smith JP, Farner DS (1982) Endocrine responses of white-crowned sparrows to environmental stress. Condor 84:399–409.

Wingfield JC, Moore MC, Farner DS (1983) Endocrine responses to inclement weather in naturally breeding populations of white-crowned sparrows (*Zonotrichia leucophrys pugetensis*). Auk 100:56–62.

Chapter 8

The Influence of Light on the Mammalian Fetus

DAVID R. WEAVER and STEVEN M. REPPERT

I. Introduction

The mammalian fetus has generally been considered to be shielded from the outside world by its presence in the womb. More recently, the presence of stimuli in the intrauterine environment capable of activating receptors for olfaction/gustation, touch, and vestibular sensation has been documented (cf. Bradley and Mistretta 1975). Prenatal development of the receptors for some senses has also been demonstrated, especially in more precocious species, and it is currently believed that the fetus perceives stimuli in several modalities.

Visual stimuli have been regarded as the least likely environmental stimuli to reach the fetus in detectable levels because of the low transmission of light through maternal tissue. While light may, in fact, reach the fetus in detectable levels, there is no evidence to suggest that the fetus responds directly to light while in the uterus. However, it would be incorrect to conclude that environmental lighting does not influence the fetus. Work in this and several other laboratories has clearly demonstrated that the environmental light–dark cycle is perceived indirectly by the fetus. During gestation, the dam acts as a transducer, communicating two aspects of the prenatal light–dark cycle to the fetus; these aspects are the time of day (the phase of the light–dark cycle) and daylength (the length of the light portion of the daily light-dark cycle). In this chapter, we will review the evidence that these two aspects of environmental lighting are perceived by the mammalian fetus and have physiological consequences for subsequent development.

II. Fetal Perception of Circadian Phase

A. Neural Substrates and Developmental Appearance of Circadian Rhythms

Perception of the timing of light–dark cycles by the fetus has been demonstrated by showing that circadian rhythms in the offspring are influenced by the prenatal light–dark cycle. Circadian rhythms are those daily rhythms which persist in the absence of time cues with a cycle length of approximately 24 hours. Under normal conditions, daily environmental cues reset or entrain the biological clock which regulates these rhythms, synchronizing the expression of physiological and behavioral rhythms to the appropriate time of the day. In mammals, light is the most potent entraining stimulus.

The system which is responsible for the generation, entrainment, and expression of circadian rhythms has been referred to as the circadian timing system. This system can be divided into three major components (Pittendrigh and Bruce 1957; Figure 8-1): a circadian pacemaker (biological clock) drives the system, input pathways entrain the pacemaker to the 24-h day, and output pathways couple the pacemaker to other neural structures for the overt expression of circadian rhythms.

Based on extensive studies in rodents, the suprachiasmatic nuclei (SCN) of the anterior hypothalamus function as such a circadian pacemaker in mammals, generating and regulating numerous circadian rhythms (for reviews, see Takahashi and Zatz 1982, Moore 1983). Input pathways relay entraining stimuli from the external environment to the circadian pacemaker in the SCN. The most important entrainment pathway is the monosynaptic retinohypothalamic pathway to the SCN (Moore and Lenn 1972, Hendrickson et al. 1972); this pathway is necessary for entrainment of circadian rhythms to environmental light–dark cycles. An indirect retinal pathway to the SCN relays in the ventral lateral geniculate nuclei (Swanson et al. 1974, Ribak and Peters 1975) and modulates the entrainment process (Albers and Ferris 1984, Albers et al. 1984, Meijer et al. 1984). Output pathways from the SCN are numerous, driving a diverse array of behavioral and hormonal rhythms. The pathway leading from the SCN to the pineal gland, regulating the synthesis of melatonin, is the most well-delineated output pathway (Tamarkin et al. 1985).

The development of a functioning circadian timing system depends on the

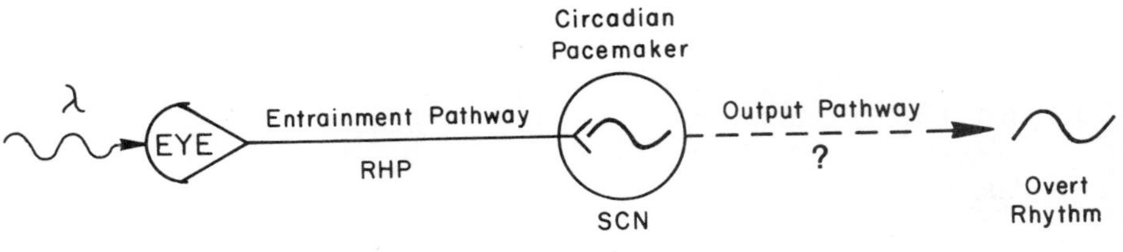

Figure 8-1. Major components of the mammalian circadian timing system. RHP, retinohypothalamic pathway; SCN, suprachiasmatic nuclei.

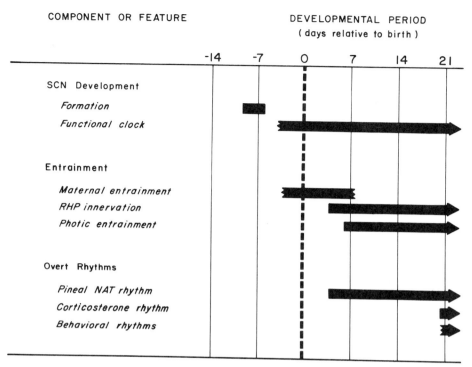

Figure 8-2. Developmental scheme of the circadian timing system of rats. The solid bars represent the time that a component or feature is present, while a jagged edge indicates that the component or feature may be present before or after that time. Behavioral rhythms refer to locomotor activity and drinking behavior. SCN, suprachiasmatic nuclei; RHP, retinohypothalamic pathway; NAT, *N*-acetyltransferase.

maturation of each of its components (Figure 8-2). In rats, the SCN undergo neurogenesis between days 13 and 16 of gestation (Ifft 1972, Altman and Bayer 1978) (ages standardized to day 0 = day of sperm positivity). Immature synapses first appear in the SCN on gestational day 18 (Koritsanszky 1981), and the vast majority of synapses appear postnatally (Lenn et al. 1977). The retinohypothalamic pathway begins innervating the SCN during the immediate postnatal period (Felong 1976, Stanfield and Cowan 1976, Terubayashi et al. 1985) and synapses of retinal fibers with SCN neurons can be identified ultrastructurally on day 6 of age (Guldner 1978).

Retina-mediated entrainment of rat pups to environmental lighting begins around day 6 (Duncan et al. 1986). Overt rhythmicity is also a postnatal event, with most behavioral and hormonal rhythms not expressed until the second or third weeks of life, possibly because of the relatively slower maturation of output pathways (for a review, see Davis 1981). Thus, in rats, the circadian timing system is not mature until well into the postnatal period.

B. Prenatal Influences on Developing Circadian Rhythmicity

1. Postnatal Paradigms

While overt circadian rhythms are not expressed until the postnatal period in rodents, the biological clock which underlies circadian rhythmicity is functioning during gestation, and prenatal influences on the timing of the fetal clock can be demonstrated. Deguchi (1975) pioneered the use of postnatal rhythmicity to show that the fetus is coordinated to the environmental light–dark cycle experienced during gestation. To demonstrate influences on the circadian timing system which occur before the overt expression of rhythmicity, he determined the phase of a rhythm monitored under constant conditions during the postnatal period. Since circadian rhythms persist under constant conditions with a cycle length close to 24 h, the phase of a rhythm in the postnatal period can be used to infer the phase of the biological clock at earlier developmental stages. For these studies, Deguchi monitored the activity of the enzyme N-acetyltransferase (NAT) in the pineal gland; pineal NAT is the first readily measurable circadian rhythm expressed by developing rats (Ellison et al. 1972). From its inception the NAT rhythm accurately reflects circadian output from the SCN (Deguchi 1982), so it is a valuable measure by which an animal's circadian time can be estimated.

Deguchi (1975) found that rat pups born and reared under constant conditions express a synchronous population rhythm in NAT activity that is in phase with the circadian time of the dam. Furthermore, by fostering pups from birth (under constant conditions) with dams whose circadian time is opposite that of the original dam, the phase of the pups' NAT rhythm shifts toward that of the foster dam. These findings suggested that a circadian clock is oscillating at or before birth and that its phase is coordinated with the dam prenatally and is also influenced by the dam postnatally.

Since Deguchi's report, several investigators have used a variety of postnatal rhythms to confirm the perinatal coordination of maternal and pup rhythms (see Table 8-1). The extent of the postnatal maternal influence varies greatly between these studies, perhaps due to strain differences or differences in the end point being measured. While some of these studies indicate that pup timing can be modified by the dam during the postnatal period, most of these studies show that there is a prenatal influence as well. It has been consistently shown that the magnitude of the postnatal maternal influence changes throughout development, being greatest during the first week of life (Takahashi and Deguchi 1983, Reppert et al. 1984). It appears that maternal influences during the postnatal period serve to maintain or reinforce the timing which has been set during the prenatal period, until the animal develops the potential for direct light–dark entrainment.

2. Prenatal Paradigms

Because of the possibility that some rhythmic aspect of the birth process itself could start or set the timing of the developing clock, postnatal paradigms cannot

Table 8-1. Studies of Perinatal Communication of Circadian Phase in Mammals[a]

Species	Measure	Influence on Pup Phase		Reference
		Prenatal	Postnatal	
Rat	NAT activity	Yes	Minor	Deguchi 1975
(Rattus	NAT activity	Yes	Yes	Deguchi 1977
norvegicus)	NAT activity	Yes	Variable	Reppert et al. 1984
	NAT activity	—	Yes	Takahashi & Deguchi 1983
	NAT activity	Yes	Yes	Reppert & Schwartz 1986a
	NAT activity	Yes	—	Reppert & Schwartz 1986b
	Drinking behavior	Yes	—	Reppert & Schwartz 1986b
	Locomotor activity	—	Yes	Takahashi et al. 1984
	Locomotor activity	—	Yes	Deguchi 1977
	Corticosterone	—	—	Hiroshige et al. 1982a
	Corticosterone	Yes	Variable	Hiroshige et al. 1982b
	Corticosterone	Yes	Variable	Hiroshige et al. 1982c
	Corticosterone	—	Yes	Takahashi et al. 1982
	Corticosterone	—	Yes	Takahashi & Deguchi 1983
	Corticosterone	—	Yes	Yamazaki & Takahashi 1983
	Corticosterone	Yes	Yes	Honma et al. 1984a
	Corticosterone	Yes	Yes	Honma et al. 1984b
	SCN metabolic activity	No	Yes	Fuchs & Moore 1980
	SCN metabolic activity	Yes-f	—	Reppert & Schwartz 1983
	SCN metabolic activity	Yes-f	—	Reppert & Schwartz 1984b
	SCN metabolic activity	Yes-f	—	Reppert & Schwartz 1986a
	SCN metabolic activity	Yes-f	—	Reppert & Schwartz 1986b
	SCN vasopressin mRNA	Yes-f	—	Reppert & Uhl 1987
Syrian hamster	Locomotor activity	—	—	Davis & Gorski 1986
(Mesocricetus	Locomotor activity	Yes	Variable	Davis & Gorski 1985
auratus)	Locomotor activity	Yes	Variable	Davis & Gorski 1983

Table 8-1. *continued*.

| Species | Measure | Influence on Pup Phase | | Reference |
		Prenatal	Postnatal	
Djungarian hamster (*Phodopus sungorus*)	Locomotor activity	—	—	Pratt 1981
Spiny mice (*Acomys cahirinus*)	Locomotor activity	Yes	No	Weaver and Reppert 1987
Field mice (*Mus booduga*)	Locomotor activity	—	—	Viswanathan & Chandrashekaran 1984
	Locomotor activity	—	Yes	Viswanathan & Chandrashekaran 1985
Squirrel monkey (*Saimiri sciureus*)	SCN metabolic activity	Yes-f	—	Reppert & Schwartz 1984a

ªEach of these studies provides evidence for within-litter coordination of circadian phase during the perinatal period. In some studies, the relative potency of pre- and postnatal influences has been assessed, allowing a comparison of their potency as indicated in the "Pre" and "Post" columns. Where either pre- or postnatal influences have not been assessed, the appropriate column contains dashes. Rhythms demonstrated in fetuses are indicated by the letter f. Abbreviations: NAT, pineal *N*-acetyltransferase activity; SCN, suprachiasmatic nuclei.

conclusively show that a circadian clock actually functions in utero. Demonstrating prenatal function of the circadian clock requires a method that can measure an intrinsic, functionally relevant property of the clock itself. A method proven useful for monitoring the oscillatory activity of the SCN in adult rats is 2-deoxyglucose (2-DG) autoradiography (Schwartz and Gainer 1977, Schwartz et al. 1980); this method allows for the in vivo determination of the rates of glucose utilization (metabolic activity) of individual brain structures. Fuchs and Moore (1980) were the first to use 2-DG autoradiography during the pre- and early postnatal periods to determine if this rhythm of metabolic activity is expressed in the SCN of neonatal rats. These investigators found that 2-DG uptake in the pup SCN was higher during the day than at night beginning on postnatal day 1, and the difference was not dependent upon light exposure on day 1. They were unable to find a day–night difference in the fetus, however. The possibility that the coordination of rhythmicity between a dam and her pups was the result of a coordinating effect of the birth process or maternal–fetal interactions on the first day of life remained until Reppert and Schwartz (1983) successfully demonstrated a circadian rhythm in metabolic activity in the fetal SCN with the 2-DG technique. This study indicated that the fetal SCN show circadian variation in metabolic activity and that this fetal oscillation is coordinated to the appropriate time of day.

Coordination of pup rhythmicity to the environment is achieved by maternal transduction of time of day information. In experiments in which pregnant dams were enucleated and placed in a lighting cycle opposite to their endogenous phase, pups were found to be coordinated to the dam and not to the external lighting cycle (Reppert and Schwartz 1983). Thus, environmental lighting does not directly influence fetal circadian phase in the rat. Instead, environmental lighting acts to entrain the dam, and through maternal–fetal communication of circadian phase the fetuses are entrained to the dam.

The day–night variation of metabolic activity in the fetal SCN is evident as early as day 19 of gestation (Reppert and Schwartz 1984b; see Figure 8-3). This early functional development is quite remarkable when compared with the SCN's morphological development (see above).

Until very recently, the 2-DG method provided the only available means of directly investigating the function of the fetal biological clock. However, it has been shown that vasopressin mRNA levels can also be used as an intrinsic marker of the oscillatory activity of the SCN during fetal life (Reppert and Uhl 1986). A population of SCN neurons contain the neuropeptide arginine vasopressin (Sofroniew and Weindl 1980), and a circadian rhythm in vasopressin levels in cerebrospinal fluid originates in the SCN (Reppert, Schwartz and Uhl 1987b). Since vasopressin mRNA levels exhibit a prominent day–night variation in adult rats (Uhl and Reppert 1986), vasopressin mRNA levels were examined throughout early development. Reppert and Uhl (1986) found a day–night oscillation in vasopressin mRNA levels in the fetal SCN beginning on day 21 of gestation, and the oscillation is in phase with the circadian time of the dam.

In summary, these studies demonstrate that the fetal biological clock is functioning late in gestation and is coordinated to environmental lighting by the dam.

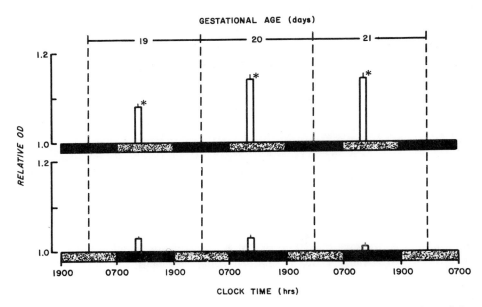

Figure 8-3. Developmental appearance of the day–night rhythm of metabolic activity in the fetal SCN. The relative optical densities (OD) of the fetal SCN (OD of SCN/OD of adjacent hypothalamus) were determined on gestational days 19, 20, and 21. The experiment was designed so that subjective-day (the period when the lights would be on had the animals remained in a light–dark cycle; upper panel) and subjective-night (lower panel) SCN glucose utilization rates were both examined at 1100 h for each gestational age. For each injection time, four to six fetuses were randomly chosen from each of two dams, and thus the vertical bar is the mean relative OD of eight to twelve fetuses (± SEM). The bars along the bottom of the figure represent the lighting cycle during pregnancy, with stippled areas indicating subjective day and solid black bars representing subjective night. A significant day–night variation was present at each gestational age (*$p < 0.001$). (From Reppert and Schwartz 1984b.)

We have already stated that the fetal rat is incapable of responding to lighting directly until the postnatal period. How then does the mother rat coordinate rhythmicity in the fetuses to the environmental light–dark cycle?

C. Mechanism of Fetal Perception of Circadian Phase

The maternal–fetal coordination of phase appears to be the result of entrainment of the fetuses to some signal from the dam. The involvement of the maternal SCN in the generation of the entraining signal(s) has been demonstrated by destroying the maternal nuclei early in gestation. These experiments suggest that the entraining signal is a circadian rhythm, and it depends on the integrity of the maternal SCN.

Using Syrian hamsters, Davis and Gorski (1983) were the first to show that complete lesions of the maternal nuclei on day 7 of gestation disrupt the timing of the developing circadian system. At the time of weaning, the phases of wheel-running behavior in pups born to and reared by SCN-lesioned dams under constant conditions were scattered throughout the 24-h day, while the rhythms from pups of sham-operated dams were coordinated with one another and with their dams. They also found that maternal SCN lesions on day 12 of the 16-day gestation period in hamsters do not disrupt fetal synchronization, suggesting that phase has already been set in the fetuses by this time. Honma et al. (1984a, 1984b) have evidence that in rats such coordination may occur by day 9 of gestation (day 10 when insemination is referred to as day 1), although several of the animals used in their studies had incomplete SCN lesions. Furthermore, their conclusion is based on a moderate (4-h) phase difference between groups reared postnatally by SCN-intact foster dams; since the corticosterone rhythm used in these studies can clearly be influenced by noncircadian factors, including feeding and stress, the results of these studies should be interpreted cautiously.

Reppert and Schwartz (1986b) utilized *three* markers of the developing circadian system in rats to examine the effects of maternal SCN lesions. The lesions were performed on day 7 of gestation; SCN glucose utilization was examined during prenatal life, while pineal NAT activity and drinking behavior were monitored during postnatal life. All animals referred to as SCN lesioned had histologically confirmed complete SCN ablation.

Maternal SCN lesions on day 7 of gestation disrupt the normal rhythmic population profile of metabolic activity in the fetal SCN. There was no day–night rhythm of fetal SCN metabolic activity following maternal SCN lesions. Instead, mean values of metabolic activity were similar during subjective day (the period of time when the lights would have been on had the animals remained in a light–dark cycle) and subjective night (the period when the lights would have been out), and the mean values were intermediate between the normally high day and low night values. Visual inspection of these autoradiographs showed that the metabolic activity of the individual fetal SCN within each litter was quite varied and spanned the inactive (invisible) to active (visible) range. This within-litter occurrence of inactive and active values was not found within litters from either intact or sham-operated dams, in which virtually all fetal SCN are either metabolically active during subjective day or inactive during subjective night.

Maternal SCN lesions on gestational day 7 also disrupt the normal rhythmic population profile of pineal NAT activity for 10-day-old pups born to SCN-lesioned dams (and reared in constant darkness). The population profile of these animals did not exhibit a significant daily rhythm, whereas a normal, rhythmic profile was exhibited by pups born to sham-operated dams and reared by SCN-lesioned dams.

Because both the DG and NAT activity rhythms were monitored from a population of animals, their disruption could indicate either loss of rhythmicity for individual animals or desynchronization of the litter while the rhythms in each animal persist. The latter interpretation is probably correct because individual rat pups born to and reared by SCN-lesioned dams and monitored under constant conditions express free-running rhythms of drinking behavior after weaning. Furthermore, the phases of drinking offset at weaning within litters from SCN-lesioned dams were not different from a random distribution (over the 24-h period), while those from sham-operated dams were synchronous with each other and with the dam (Figure 8-4). As noted earlier, Davis and Gorski (1983) came to a similar conclusion by examining locomotor activity in pups from dams which underwent SCN lesions on gestational day 7.

The maternal SCN are also necessary for postnatal communication of phase (Reppert and Schwartz 1986b). Pup NAT activity profiles are rhythmic when pups are reared by SCN-lesioned dams either pre- or postnatally but are not rhythmic when they are reared by SCN-lesioned dams during *both* the pre- and postnatal periods (Figure 8-5). The synchronization found with maternal entrainment during only the postnatal period (pups born to SCN-lesioned dams and reared by intact dams) suggests that one must be extremely cautious when using only postnatal rhythms to examine prenatal circadian influences.

Since SCN lesions disrupt virtually all circadian rhythms in rats and hamsters, the results of the lesion studies do not provide much information regarding the identity of the maternal entraining signal. They do provide a standard which is quite useful in assessing the ability of other manipulations of the dam to disrupt phase communication.

A number of studies have been conducted to try to identify the signal which communicates phase to the rat fetus. Since the in utero environment provides a rich source of hormonal signals from the mother, studies in rats have focused on the possibility that the maternal signal for entrainment is hormonal. A conceptual model for maternal–fetal communication of phase, which shows the involvement of the maternal SCN and the generation of a maternal signal, is presented in Figure 8-6.

If a rhythm in a certain hormone is the signal for maternal–fetal communication of phase, then removal of that hormone (by removal of its source) should prevent entrainment of the fetuses to the dam. Therefore, the effect of removal of a number of endocrine glands (in separate experiments) on maternal–fetal entrainment has been examined.

There is a robust rhythm in melatonin in the maternal circulation, and since melatonin can cross the placenta (Klein 1972), a melatonin rhythm in the dam is reflected in the fetal circulation (Reppert et al. 1979). Because of these factors, melatonin was a prime candidate for the maternal signal. Rhythms in melatonin were prevented by removing the maternal pineal gland on gestational day 0; pinealectomy of Zivic-Miller rats by the supplier (as done in this study) eliminates the circulating melatonin rhythm, as well as measurable levels of the hormone (Lewy et al. 1980).

The results of maternal pinealectomy on the metabolic activity (2-DG) rhythm in the fetal SCN were enigmatic (Reppert and Schwartz 1986a). As occurs when maternal–fetal communication exists, all fetal SCN were metabolically inactive during subjective night. During subjective day, all fetal SCN from one of the pinealectomized dams were metabolically active (normal), but all fetal SCN from the other dam were metabolically inactive. This experiment was repeated; again all fetal SCN from one of the two subjective-day dams were metabolically inactive. This pattern does not reproduce the pattern found after maternal SCN lesions, where metabolic activity levels of the fetuses within individual litters are scattered (see above). The significance of the 2-DG data concerning maternal

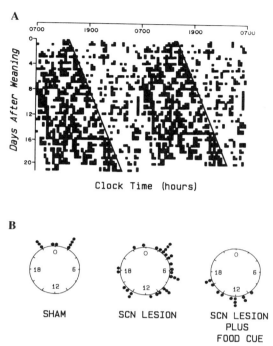

Figure 8-4. (A) Temporal profile of drinking behavior from a rat pup born and reared in constant darkness. On postnatal day 21, the pup and its littermates were weaned, placed into individual cages, and drinking behavior was monitored continuously for the next 21 days. The data have been transformed in a semiquantitative way in which the number of drinking bouts per 30-minute time bin is represented by the number of boxes (from 0 to 4) stacked vertically in the time bin. An eye-fitted line has been drawn through the daily offset times of drinking for the 21-day period, and the intercept at weaning was determined. The record is double plotted to facilitate visual inspection. (B) Phases of drinking offset from pups of either sham-operated dams, SCN-lesioned dams, or SCN-lesioned dams given morning food cue. Dams were reared in diurnal lightning (lights on at 0700–1900) during pregnancy and surgery was performed on gestational day 7. Two days before birth, the dams were placed in constant darkness, and the pups were born and reared in constant darkness. After weaning, the pups were monitored for drinking behavior and the phase of drinking offset was determined for each pup on day 21 as described for panel A. The data are presented in a modified version of the method of Davis and Gorski (1985). Each of the three large circles represents 24 hours of day 21, and each small solid circle represents the time of drinking offset of one pup. Pups from two litters from sham-operated dams, five litters from lesioned dams, and three litters from lesioned dams given food cue have been pooled by the dam's treatment. To combine litters in this way, the mean phase of pups from sham-operated dams was determined (time zero) and the phase of pups in other groups is expressed relative to that reference point.

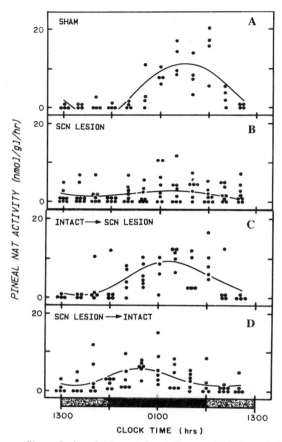

Figure 8-5. Daily profiles of pineal *N*-acetyltransferase (NAT) activity for 10-day-old pups (A) born to and reared by sham-operated dams, (B) born to and reared by SCN-lesioned dams, (C) born to intact dams and reared by SCN-lesioned dams, or (D) born to SCN-lesioned dams and reared by intact dams. SCN lesions and the sham surgery were performed on gestational day 7. Two days before birth, the dams were transferred from diurnal lighting (12L:12D) into constant darkness. The pups were sacrificed in constant darkness over a 24-h period on postnatal day 10. Each point is the NAT value for a single pup. The waveforms were derived from sine-wave analysis of the profiles and are for illustrative purposes only. The prenatal light–dark cycle is depicted along the abscissa (solid bar represents night, open bar represents day). Pup NAT activity profiles were rhythmic when reared by an SCN-lesioned dam either pre- or postnatally but were not rhythmic when reared by SCN-lesioned dams during both the pre- and postnatal periods. (Modified from Reppert et al. 1984.)

pinealectomy is unclear. It appears that there is still within-litter coordination, but the coordination of fetal rhymicity to the dam and the environment may be affected. It is possible that pinealectomy has some subtle influence on the phase of the fetal rhythm, without affecting rhythmicity or maternal–fetal co-ordination per se.

Figure 8-6. Conceptual model of maternal–fetal communication of circadian phase. Light-induced neural signals are conveyed to the dam's SCN by her retinohypothalamic pathway (RHP), entraining her circadian rhythms. A maternal output signal then entrains the fetal clock at a time when the innervation of the fetal SCN by the RHP is incomplete.

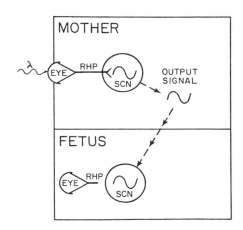

Because of this unusual 2-DG pattern after maternal pinealectomy, Reppert and Schwartz (1986a) examined another marker of the developing circadian system, the rhythm of pineal NAT activity in 10-day-old pups. Pups born to pinealectomized dams were reared from birth in constant darkness by SCN-lesioned foster dams; as presented above, the postnatal maternal influence is abolished by complete lesions of the maternal SCN, so that this paradigm allows assessment of prenatal influences on phase while preventing complicating postnatal influences. Pups born to either intact or pinealectomized dams and reared by SCN-lesioned dams throughout the postnatal period has significant daily rhythms of NAT activity, and the phase of the rhythms for the two groups was similar. Thus, maternal pinealectomy does not abolish maternal coordination of fetal rhythmicity as monitored during the postnatal period with pineal NAT activity. Thus, despite the unusual results obtained with 2-DG, the maternal pineal gland does not appear to be necessary for maternal–fetal communication of phase.

Removal of the maternal adrenals, thyroid-parathyroids, pituitary, or ovaries in separate experiments has been used to assess whether these glands are necessary for maternal–fetal communication of phase. None of these extirpations abolish the clear day–night rhythm of metabolic activity in the fetal SCN (Reppert and Schwartz 1986a) when performed on or before day 7 of gestation (the time when maternal SCN lesions are disruptive). Furthermore, the maternal eyes, a potential source of both neural and endocrine signals, are not necessary for the communication since blind dams (enucleated on day 2 of gestation) synchronize the circadian clocks of their fetuses (Reppert & Schwartz 1983).

Following maternal SCN lesions on day 7 of gestation, disruption of maternal circadian rhythms and disruption of maternal–fetal coordination occur together. One approach to determining which aspect of maternal rhythmicity is involved in fetal entrainment is to artificially restore rhythmicity in a dam which has no endogenous rhythmicity. There is a circadian rhythm in food consumption, and restricted access to food in SCN-lesioned rats produces rhythms in locomotor activity and temperature (Krieger et al. 1977, Stephan 1981). We have used a similar food access restriction paradigm (food cue) in SCN-lesioned dams in order to artificially impose daily rhythmicity in these animals. Our preliminary data suggest that the rhythmic ingestion of food can entrain fetuses of SCN-lesioned dams. For this study (Reppert and Weaver, unpublished), dams were bred and reared in diurnal lighting. On day 7 of gestation, they received bilateral SCN lesions and food access was restricted to 4 h per day during the early light period; the time of food access in these animals was forced to be opposite the time when it would normally occur in shams with *ad libitum* food. On day 19, food was made freely available and the animals were placed in constant darkness.

The phases of drinking offsets at weaning within litters from SCN-lesioned dams given food cue were relatively synchronized, in contrast to SCN-lesioned dams not given food cue (Figure 8-4). Furthermore, the phases of drinking offsets for these litters were opposite those for litters born to sham-operated dams that were not food restricted. Thus, rhythmic food ingestion during pregnancy appears to synchronize developing circadian phase. This finding provides a significant advance in our investigations of the maternal output signal(s) communicating phase to the fetus. Rhythmic food ingestion clearly would cause rhythmic fluctuations in nutrient levels in the blood, and this would in turn induce physiological responses to the presence of nutrients. In SCN-lesioned dams, food restriction may cause generation of a nutrient-related signal which entrains the fetuses in a manner parallel to that which occurs in intact dams. Alternatively, restricted feeding of SCN-lesioned dams may lead to the generation of other rhythms (e.g., temperature or activity) which are involved in setting the timing of the pups. We are currently investigating these possibilities.

D. Possible Functions of Maternal–Fetal Communication of Circadian Phase

The widespread presence of perinatal communication of circadian phase in a number of mammalian species indicates that this phenomenon is of considerable adaptive value. Maternal entrainment would presumably coordinate the pups' circadian phase to that of the dam and to the environment, so that when physiological and behavioral rhythmicity later develops, these rhythms are expressed in proper relationship to one another and to the 24-hour day, allowing the young animal to assume its temporal niche. In burrow-dwelling mammals, the offspring are probably not directly exposed to the daily light—dark cycle until they can crawl out of the burrow. Using a simulated burrow environment, Pratt (1981) found that Djungarian hamster pups do indeed remain in the burrow (and thus in constant darkness) for the first two weeks of life. When pups begin emerging, their behavioral rhythms are already attuned to the light–dark cycle outside, indicating that the developing circadian system has been entrained by the dam. If the pups were not coordinated to the environment by the mother, they might exhibit inappropriate timing of emergence from the burrow or disadvantageous behavioral patterns when they emerged. Further, if each pup were to develop with its own rhythmicity, there would be a period of disorganization of the litter that could be detrimental to some members of the litter. For example, coordination of a dam's willingness to nurse and of all pups in the litter to suckle would obviously be of benefit. Circadian disorganization at the level of the individual or litter could thus threaten survival.

Another possible function of the developing clock and its entrainment during fetal life is that it might be involved in the initiation of parturition. The time of day of birth is gated to the daily light–dark cycle by a circadian mechanism in a number of species (Kaiser and Halberg 1962, Rossendale and Short 1967, Mitchell and Yochim 1970, Jolly 1972, Lincoln and Porter 1976). In rats, for example, births are more prevalent during the day, and the timing of births (relative to the time from conception) can be manipulated over a 36-hour "window" by altering the phase of the prenatal light–dark cycle (Reppert et al. 1987a). If the fetuses of polytocous mammals, such as rats, are involved in the initiation of birth, then all the fetal-placental units would have to function in unison to initiate this process. Maternal entrainment of the circadian clock of each fetus may be important for providing this synchrony and, therefore, may be important for the initiation of birth.

III. Fetal Perception of Daylength

In addition to fetal perception of time of day, the length of the light portion of the environmental light–dark cycle is also perceived by the fetus in some species. This influence of prenatal lighting has been demonstrated in two species in which daylength regulates reproductive maturation in juveniles.

A. Photoperiodic Regulation of Reproduction

Seasonal variations in environmental conditions occur throughout temperate regions, and many animals undergo seasonal physiological changes. Many species prepare for the severe climate and reduced food availability characteristic of winter by increasing stores of body fat, growing a thicker but lighter-colored coat, and, in some species, entering hibernation or daily torpor. Most importantly, mating is restricted to portions of the year such that young are born in the spring and summer, when conditions are most favorable for their survival. Daylength (e.g., the length of the light period of the daily light–dark cycle) is probably the most reliable and thus most widely used indicator of season.

In adult Syrian and Djungarian hamsters, long daylengths stimulate or maintain reproductive function while short photoperiods induce testicular regression in males and a cessation of estrous cyclicity in females (for a review, see Tamarkin et al. 1985). In some photoperiodic species, short daylengths which suppress reproductive function in adults also suppress gonadotropins and prevent testicular growth in juvenile males (Pinter 1968, Hoffmann 1978, Carter and Goldman 1983, Horton 1984a,b, Yellon and Goldman 1984). It appears that the response to a particular daylength is not absolute; rather, the response to a photoperiod depends in part on the preceding photoperiodic experience

(Hoffmann 1984). In the montane vole and Djungarian hamster, the photoperiodic regulation of puberty allows examination of the development of "photoperiodic history"; these studies indicate that an animal's photoperiodic history begins prenatally.

B. Prenatal Communication of Daylength

Evidence for a prenatal influence of photoperiod on postnatal reproductive maturation was first provided by Negus et al. (1977), who observed the appearance of juvenile montane voles (*Microtus montanus*) with inhibited reproductive development shortly after the summer solstice in a natural population. The photoperiod at this time is long and is clearly stimulatory to reproduction in adults. Of more relevance to the juveniles, however, is the fact that the daylength is decreasing. Horton (1984b) subsequently showed in a laboratory population of montane voles that reproductive maturation could be altered by manipulations of the prenatal photoperiod. The reproductive response of the male offspring was not determined by the postnatal photoperiod alone. Instead, both pre- and postnatal daylengths influenced development. Male pups which were shifted on the day of birth from a long photoperiod (16L:8D) to a shorter one (14L:10D) had inhibited reproductive development, while males reared in this shorter photoperiod both before and after birth had stimulated reproductive development. A similar finding has recently been reported by Stetson et al. (1986) using Djungarian hamsters. Cross-foster studies by Horton (1985) and Goldman (Reppert et al. 1985) indicate that the inhibition of testicular development following a shift to a shorter daylength on the day of birth is not dependent upon shift-induced alterations in maternal physiology; rather, a prenatal signal communicating photoperiod is received by the fetus.

Because of our interest in maternal–fetal communication of light information, we set out to study the mechanism of maternal–fetal communication of daylength in Djungarian hamsters (Weaver and Reppert 1986). We confirmed that prenatal photoperiod influences postnatal reproductive development in our laboratory by raising litters of Djungarian hamsters under one of three conditions. Some litters were shifted from a prenatal photoperiod of 16L:8D (16L) to 14L:10D (14L) on the day of birth (16L-14L group). Other groups of litters were reared in 14L both pre- and postnatally (14L-14L group) or 16L both pre- and postnatally (16L-16L group). Males that experienced a 2-h decrease in daylength on the day of birth (16L-14L group) had significantly reduced absolute and relative testicular weights on day 34 compared to the 14L-14L groups (see Table 8-2); both males and females of the 16L-14L group also had significantly depressed body weights compared to 14L-14L animals. The presence of these differences between groups reared in the same postnatal photoperiod indicates that the prenatal photoperiod influences somatic and reproductive development. The possibility that a prenatal daylength of 16L is somehow harmful to postnatal reproductive development is excluded because the 16L-16L group had rapid gonadal development.

Table 8-2. Effects of Photoperiod, Maternal Pinealectomy, and Prenatal Melatonin Infusion on Reproductive Development in Juvenile Male Djungarian Hamsters

Dam, photoperiods[a]	n	Testicular weight	Body weight
Intact dam, 16L-16L	25	457 ± 18	23.0 ± 0.5
Intact dam, 16L-14L	35	69 ± 6	19.9 ± 0.4
Intact dam, 14L-14L	36	282 ± 19	22.7 ± 0.4
PNX dam, 16L-14L	17	60 ± 8	19.6 ± 0.7
PNX dam, 14L-14L	17	71 ± 7	20.5 ± 0.4
PNX dam, 14L-14L + Mel-6	11	57 ± 11	20.7 ± 0.7
PNX dam, 14L-14L + Mel-8	8	303 ± 30	25.2 ± 0.5
PNX dam, 14L-14L + Mel-8 D	9	265 ± 71	25.3 ± 1.0

[a]Intact (unoperated) and pinealectomized (PNX) dams reared offspring in various pre- and postnatal photoperiods (e.g., 16L-16L indicates pre- and postnatal photoperiods of 16L:8D). Some PNX dams received melatonin infusions during gestation; 50 ng melatonin in 0.2 ml saline (0.025% ethanol) was delivered over either 6 or 8 h (Mel-6 and Mel-8, respectively, with the midpoint of infusion at the midpoint of the dark phase of the 14L:10D light–dark cycle. In a third infusion group, 8-h melatonin infusions were delivered during the day (Mel-8 D), with the midpoint of infusion coincident with the midpoint fo the light portion of the 14L:10D light–dark cycle. (Data of Weaver and Reppert, 1986.)

The ability to detect changes in photoperiod, and therefore the photoperiodic regulation of reproduction, is blocked by pinealectomy in those species in which it has been examined (for reviews, see Goldman and Darrow 1983, Karsch et al. 1984, Tamarkin et al. 1985). Studies using exogenously administered melatonin have indicated that this hormone is the active pineal factor mediating these effects (e.g., Tamarkin et al. 1977, Goldman et al. 1982, Bittman and Karsch 1984). In several species, the duration of melatonin secretion is proportional to the length of the night (Goldman et al. 1982, Bittman et al. 1983, Florant and Tamarkin 1984, Darrow and Goldman 1985), and the duration of melatonin elevation appears to be the critical parameter in mediating its effects on reproductive status in Djungarian hamsters and ewes (Carter and Goldman 1983, Bittman and Karsch 1984).

We examined the role of the maternal pineal gland in prenatal communication of daylength by assessing the effect of removal of the maternal pineal gland. In contrast to the results from intact dams, there were no differences in testicular development or body weight between 16L-14L and 14L-14L offspring of pinealectomized dams (see Table 8-2). Similarly, Elliott and Goldman (1986) recently reported that maternal pinealectomy prevents the influence of prenatal daylength on testicular weight.

We tested the hypothesis that maternal melatonin is the pineal product communicating daylength to the fetus by infusing the hormone during the last 3 to

10 days of gestation into pinealectomized dams maintained in 14L. The method for chronic subcutaneous infusion of melatonin was similar to that described by Carter and Goldman (1983). Melatonin (50 ng) was delivered over 6 or 8 h per night; these durations were chosen because 8-h infusions into pinealectomized juvenile male Djungarian hamsters simulate the effect of exposure of a pineal-intact juvenile to 14L, while a 6-h infusion simulates 16L (Carter and Goldman 1983). All offspring were reared in 14L postnatally.

Pups whose dams received 8-h infusions during gestation had stimulated testicular development on day 34, comparable to 14L-14L males from intact dams (see Table 8-2). In contrast, pups from dams receiving 6-h infusions had suppressed development, comparable to 16L-14L males from intact dams or pups of uninfused pinealectomized dams. There was no clear relationship between the number of days of infusion and the reproductive response of the offspring, and there did not appear to be a difference between 8-h infusions delivered during the day and at night. These results strongly implicate melatonin as the pineal output communicating daylength to the fetus.

Watson-Whitmyre and Stetson (this volume, Chapter 11) have also recently shown that injections of melatonin into pregnant dams which are timed to extend the endogenous melatonin elevation can mimic the effect of short days during the prenatal period on the reproductive development of the offspring.

While maternal melatonin is clearly involved in maternal–fetal communication of daylength, the mechanism by which it does so remains to be determined. Based on studies in rats, a maternal melatonin rhythm would be reflected in the fetuses (Klein 1972, Reppert et al. 1979), and melatonin could act directly in the fetus to communicate daylength. Alternatively, the duration of melatonin elevation in the dam could regulate the production of a secondary signal (or a cascade of hormonal events) which results in some other signal actually reaching the fetus. Thus, while the duration of maternal melatonin is clearly involved in maternal–fetal communication of daylength, melatonin may not be the signal which ultimately reaches the fetus. Further work is necessary to distinguish between these possibilities.

C. Physiological Significance of Prenatal Communication of Daylength

Maternal–fetal communication of daylength would be of considerable value in a long-day breeder such as the Djungarian hamster. If offspring born early in the breeding season are to reproduce in the same year in which they are born, rapid reproductive development is necessary. Late-season offspring would be at a disadvantage, however, if they were to expend their energy rapidly reaching puberty when the end of the breeding season is approaching. Clearly, no single strategy would be appropriate for all offspring. Matching the rate of development to the environmental conditions which will occur in the future would be valuable, and measuring daylength can be used to determine the season. However, any daylength occurs twice in a year, making the daylength at a single point in time useless in determining the time of year. A mechanism for comparison of the daylength at two time points is needed to determine the direction of change in daylength, and thus the season. Maternal–fetal communication of daylength provides the fetus with one daylength measurement. Upon measuring daylength on its own during the postnatal period [beginning around postnatal day 15 (Tamarkin et al. 1980, Pratt 1981, Yellon et al. 1985)], the offspring can rapidly determine the direction of change in the daylength. In this way, maternal–fetal communication of daylength allows the pups to rapidly make the seasonally appropriate choice between preparing for the summer breeding season or preparing for the winter (e.g., by altering energy metabolism and body fat stores).

IV. Possible Direct Effects of Light on Fetal Entrainment

A. Direct Versus Indirect Perception of Light in Precocious Species

To this point, we have emphasized maternal transduction of the light–dark cycle into information which is then perceived by the fetus. Maternal transduction appears to be the only mechanism for fetal perception of environmental lighting in altricial species such as rats and hamsters, in which the neural mechanisms for direct responsiveness to environmental lighting develop postnatally. For example, the neonatal retina is immature and unresponsive in rats, mice, cats, rabbits, and dogs (for a review, see Bradley and Mistretta 1975). In more precocious species, however, the capacity for retinal photoreception appears to be essentially complete at birth (e.g., humans and rhesus monkeys; for a review, see Bradley and Mistretta 1975). Since stimulation of the optic nerves can evoke neural (cortical) responses from fetal guinea pigs (Sedlacek 1971) and sheep (Persson and Stenberg 1972) when the fetus is externalized, the pathways necessary for detection of light are at least partly functional before birth in these species. It seems likely that neural connections relaying light to the circadian timing system may also be developed before birth, although the developmental appearance of connections in the primary visual system and the circadian timing system may differ. Further work is needed to determine the anatomical and functional development of the retinohypothalamic tract in precocious species. For example, this entrainment pathway is present on the day of birth in guinea pigs (Reppert, unpublished), and it would be of interest to determine if it is present prenatally.

In addition to the developmental status of the fetal nervous system, the amount of light available to the receptive tissues must be considered when assessing the potential for the direct perception of light by the fetus. Light must be present in sufficient quantities at the appropriate wavelengths if it is to be detected by the fetus. We have recently demonstrated that a significant fraction of incident light is transmitted into the uterus of pregnant rats and guinea pigs. In the blue-green range of the spectrum (450–550 nm), which is the range for maximal sensitivity to the entraining influence of light, approximately 2% of incident light reaches the lumen of the uterus (Jacques, Weaver and Reppert 1987; Figure 8-7). While we have not performed naturalistic studies to determine how much light the dams are exposed to during pregnancy, it seems likely that at least in some species the dam may expose herself to sufficient lighting to

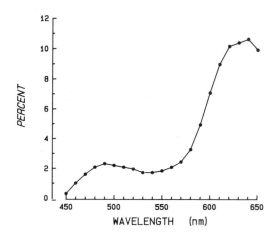

Figure 8-7. Percent of incident light transmitted into the uterus of pregnant guinea pigs, by wavelength. Light intensities were determined with a sensor probe which was unobstructed (100%) and then with the probe placed in the uterus of an anesthetized Hartley strain (albino) guinea pig on day 51 of gestation. All measurements were made with a bank of fluorescent bulbs above and to the sides of the animal; the probe remained in a fixed position relative to the source for all measurements. The figure represents the average transmission through dorsal and ventral surfaces for three dams. (Modified from Jacques et al. 1987.)

influence the fetus. The abdomens of diurnal primates would clearly be exposed to a large amount of light, and the fetal SCN of at least one primate species show a day–night variation in metabolic activity (Reppert and Schwartz 1984a). Whether this rhythm is due to maternal–fetal communication of phase or direct photic entrainment of the fetus remains to be established.

We have recently developed a model for examination of the potential for direct photic entrainment of the fetus in precocious species using the spiny mouse, *Acomys cahirinus*. Spiny mice have a gestation length of 38 to 40 days; the eyes open and the pups can move about the cage within hours of birth, and auditory and olfactory senses are functional at this time (Ruch 1967, cited by Porter and Ruttle 1975). Because of these qualities, spiny mice have been intensively studied with regard to their development of olfactory preferences and mother–offspring interactions (e.g., Porter and Etscorn 1974, Porter et al. 1980, 1983). Spiny mice are also ideal for our studies, as they can be weaned as early as day 7 and express robust circadian rhythms in running wheel behavior at weaning. The spiny mouse dam does not build a nest, and the pups are mobile shortly after birth; there would be little need for postnatal maternal influences on pup phase, as the pups are exposed to the environment (and presumably responsive to it) during the immediate postnatal period.

To validate the use of a postnatal behavioral model in assessing prenatal influences, we demonstrated that communication of phase occurs prenatally in this species and that postnatal maternal influences do not obscure prenatal entrainment (Weaver and Reppert 1987). We maintained harems of spiny mice in 14L:10D lighting, with one set of harems in diurnal lighting (LD) and another in reversed lighting (DL). Pregnant dams were placed in individual cages in constant darkness several days prior to giving birth and running wheel activity was monitored. Pups were weaned to their own running wheel cages on day 13 of life and monitored for 2 to 3 weeks. The phases of activity onset for dams and pups were determined by drawing an eye-fitted line through the daily activity onsets of each animal (Figure 8-8). On the day of weaning, the activity onset of pups differed from that of their dams by an average of 6 h in both LD and DL groups. When the phase plots were extrapolated back to the day of birth, the average dam–pup difference in activity onset was reduced to 2.5 h (Figure 8-9), indicating that the dams and their pups were significantly more coordinated on the day of birth than at weaning. The average dam–pup difference did not differ between the DL and LD litters at either day 0 or day 13, indicating that the coordination was due to the lighting cycle experienced by the dam, and not to the absolute time of day.

To assess postnatal influences on pup rhythmicity, we fostered two pups from a DL dam to an LD dam on the day of birth. When the phases of the pups were extrapolated back to the day of birth, the pups were found to be in phase with their original dam (Figure 8-10). If the foster dam had exerted a significant influence on the timing of rhythmicity in the pups, the extrapolation of the phase of the pups would not indicate such close coordination between the pups and their original dam. As previously suggested by the relative independence of the cycle length of the maternal and pup rhythms during the postnatal period in normal litters, the results of this foster study indicate that there is little or no postnatal maternal influence on pup rhythmicity in this species, making it a valuable model with which to examine prenatal influences on developing circadian phase.

Two possibilities for the mechanism for this prenatal coordination between dam and pups seemed likely: either the dam mediates the effect of the lighting cycle and entrains the pup (as in rats) or light entrains the fetus directly. To address this issue, we enucleated pregnant dams 2 to 3 weeks before birth and placed them into a lighting cycle opposite their circadian time. As in the first experiment, the blind dams were shifted to constant darkness before giving birth, and the activity of pups and their dams was monitored. Pups exposed during gestation to light out of phase with the maternal rhythmicity were coordinated with the dam at birth, and not with the lighting cycle, indicating that prenatal entrainment is maternally mediated.

While this study suggests that the dam and not the LD cycle entrains the fetuses, two features of the methods used deserve comment. First, light was given throughout the latter part of gestation, but the animals were placed in constant darkness from 5 to 8 days before birth. Thus, light may not have been available at a time when the fetus could actually respond to light (e.g., the retinohypothalamic pathway may not be in place until the last few days of gestation). Second, this study did not rule out the *potential* influence of light of higher intensity on the fetus. This study demonstrated that maternal–fetal communication of phase occurs in this species; it is also possible that light could influence the timing of rhythms in the spiny mouse fetus if provided in sufficient intensity and at the maturationally proper time of gestation. We intend to con-

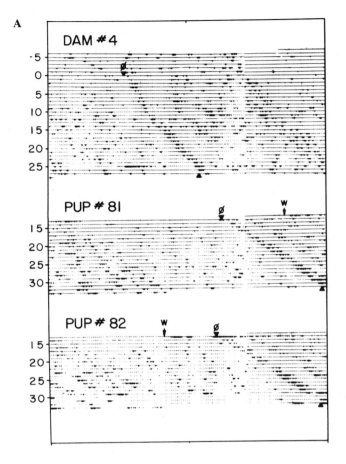

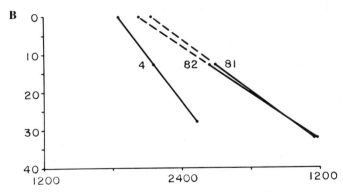

Figure 8-8. (A) Activity records of a spiny mouse dam and her two pups recorded in constant darkness. The dam was moved from 14L:10D lighting (lights on 0400–1800) to constant darkness six days before birth. The pups were weaned (W) to individual running wheel cages at two different times on postnatal day 13. For each animal, each horizontal

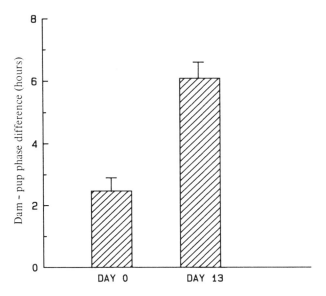

Figure 8-9. Average difference between the phase of activity onset for spiny dams and their pups on day 0 (birth) and day 13 (weaning). The times of activity onset at birth (extrapolated) and weaning were determined as shown in Figure 8-8, panel A for 18 pups from 8 dams. Error bars represent standard errors of the mean.

tinue examining the potential influence of light on the development of circadian rhythmicity, including studies of the ontogeny of the retinohypothalamic tract and the appearance of direct photic entrainment in this species.

B. Possible Effects of Direct Prenatal Detection of Light

In addition to transmitting phase to the circadian timing system of the developing fetus, directly perceived light may have other functions. Considering the permanent influences of visual deprivation on neuronal connections and function during postnatal critical periods (e.g., Hubel and Weisel 1970), the potential

---

line of the activity record represents one 24-h period from a channel of an Esterline-Angus event recorder, with successive days aligned below the first for each record. Along each horizontal line, running wheel activity is indicated by deflection of the pen from the baseline. For each animal, the phase of activity onset (ϕ) was determined by drawing an eye-fitted line through the daily activity onsets; the two triangles in each animal's record indicate two points on the line used to define activity onset. (B) Superimposition of the lines indicating activity onset (indicated by the triangles in panel A) for the dam and her pups shows that the activity onsets of the pups differ from the onset of the dam by approximately 6 hours at weaning (day 13). Extrapolation of the line representing activity onset to the day of birth (dashed lines) indicates the dam–pup difference is lower on the day of birth.

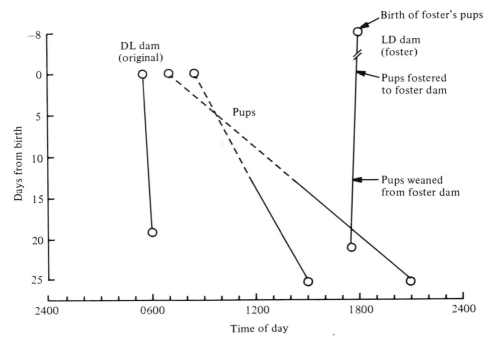

Figure 8-10. Assessment of postnatal influences on pup rhythmicity in spiny mice. Two pups born in constant darkness to a dam which had been entrained to reversed lighting during pregnancy (DL; original dam) were fostered to a dam which had been entrained to diurnal lighting during pregnancy (LD; foster dam) on the day of birth. The foster dam had given birth to two pups 8 days earlier; these pups were weaned from the dam at the time the other pups were fostered to her. The fostered pups were weaned from the foster dam on day 13, and the activity onsets of the pups and both the original and foster dams were determined as illustrated in Figure 8-8. The activity onsets of the pups were 1.5 and 2.5 h from the onset of their original dam, which is similar to the difference in onset seen in pups which are not fostered. If the foster dam had exerted a significant influence on the circadian phase of the pups, activity onset would have been affected in such a way that the line indicating activity onset would not be so close to the original dam when extrapolated back to the day of birth.

importance of photic stimulation during the prenatal period should not be ig-nored. Stimulation of visual receptors during the prenatal period may influence the development and future function of the receptors themselves as well as of neuronal elements with which they communicate (Bradley and Mistretta 1975).

V. Summary and Conclusions

We have discussed the evidence that the mammalian fetus is influenced by the prenatal light–dark cycle. Both the timing and length of daylight are commu-nicated to the fetus by the dam, influencing the timing of circadian rhythmicity and the rate of reproductive maturation, respectively. The mechanism of maternal–fetal communication of circadian phase in rats remains to be determined, but the necessity of the maternal SCN for transduction of environmental lighting into a form which can be perceived by the fetus has been demonstrated. Maternal melatonin is involved in the transfer of daylength information to the fetus, although the nature of this involvement remains to be determined. These novel forms of maternal–fetal communication are likely to have physiological relevance for the developing mammal, allowing the neonate to predict alterations in the environment which are due to the daily and annual solar cycles. As a result of maternal transduction of lighting information, the fetus can be influenced by environmental light–dark cycles while the mechanisms for direct perception of light are still immature. In precocious species, development of the neural substrates for light detection appears to occur prenatally, and the possibility of direct intrauterine perception of light by the fetus deserves further study.

Acknowledgments. We thank Drs. William J. Schwartz, Steven Jacques, and George Uhl for their collaborative efforts. This work was supported by PHS grant HD14427 and March of Dimes Basic Research Grant #I-945. S.M.R. is an Established Investigator of the American Heart Association.

References

Albers HE, Ferris CF (1984) Neuropeptide Y: role in light–dark entrainment of hamster circadian rhythms. Neurosci Lett 50:163–168.

Albers HE, Ferris CF, Leeman SE, Goldman BD (1984) Avian pancreatic polypeptide phase shifts hamster circadian rhythms when microinjected into the suprachiasmatic region. Science 223:833–835.

Altman J, Bayer SA (1978) Development of the diencephalon in the rat. I. Autoradiographic study of the time of origin and settling patterns of neurons of the hypothalamus. J Comp Neurol 182:945–972.

Bittman EL, Karsch FJ (1984) Nightly duration of melatonin secretion determines the reproductive response to inhibitory day length in the ewe. Biol Reprod 30:585–593.

Bittman EL, Dempsey RJ, Karsch FJ (1983) Pineal melatonin secretion drives the reproductive response to daylength in the ewe. Endocrinology 113:2276–2283.

Bradley RM, Mistretta CM (1975) Fetal sensory receptors. Physiol Rev 55:352–382.

Carter DS, Goldman BD (1983) Antigonadal effects of timed melatonin infusion in pinealectomized male Djungarian hamsters (*Phodopus sungorus sungorus*): duration is the critical parameter. Endocrinology 113:1261–1267.

Darrow JM, Goldman BD (1985) Circadian regulation of pineal melatonin and reproduction in the Djungarian hamster. J Biol Rhythms 1:39–54.

Davis FC (1981) Ontogeny of circadian rhythms. In: Aschoff J (ed) Handbook of Behavioral Neurobiology, Vol. 4, Biological Rhythms. Plenum Press, New York, pp. 257–274.

Davis FC, Gorski RA (1983) Entrainment of circadian rhythms in utero: role of the maternal suprachiasmatic nucleus. Soc Neurosci Abs 9:625.

Davis FC, Gorski RA (1985) Development of hamster circadian rhythms: prenatal entrainment of the pacemaker. J Biol Rhythms 1:77–89.

Davis FC, Gorski RA (1986) Development of hamster circadian rhythms: I. Within litter synchrony of mother and pup activity rhythms at weaning. Biol Reprod 33:335–362.

Deguchi T (1975) Ontogenesis of a biological clock for serotonin: acetyl coenzyme A acetyltransferase in the pineal gland of rat. Proc Natl Acad Sci USA 72:2814–2418.

Deguchi T (1977) Circadian rhythms of enzyme and running activity under ultradian lighting schedule. Am J Physiol 232:E375–E381.

Deguchi T (1982) Sympathetic regulation of circadian rhythm of serotonin N-acetyltransferase activity in pineal gland of infant rat. J Neurochem 38:797–802.

Duncan MJ, Banister MJ, Reppert SM (1986) Developmental appearance of light–dark entrainment in the rat. Brain Res 369:326–330.

Elliott JA, Goldman BD (1986) Pineal gland of pregnant Djungarian hamsters mediates reception of photoperiodic information by the developing fetus. Biol Reprod 34:221.

Ellison N, Weller JL, Klein DC (1972) Development of a circadian rhythm in pineal N-acetyltransferase. J Neurochem 19:1335–1341.

Felong M (1976) Development of the retinohypothalamic projection in the rat. Anat Rec 184:400–401.

Florant GL, Tamarkin L (1984) Plasma melatonin rhythms in euthermic marmots (*Marmota flaviventris*). Biol Reprod 30:332–337.

Fuchs JL, Moore RY (1980) Development of circadian rhythmicity and light responsiveness in the rat suprachiasmatic nucleus: a study using the 2-deoxyl[1-14C]glucose method. Proc Natl Acad Sci USA 77:1204–1208.

Goldman BD, Darrow JM (1983) The pineal gland and mammalian photoperiodism. Neuroendocrinology 37:386–396.

Goldman BD, Carter DS, Hall VD, Roychoudhury P, Yellon SM (1982) Physiology of melatonin in three hamster species. In: Klein DC (ed) Melatonin Rhythm Generating System—Developmental Aspects. Karger, Basel, pp. 201–231.

Guldner FH (1978) Synapses of optic nerve afferents in the rat suprachiasmatic nucleus. I. Identification, qualitative description, development and distribution. Cell Tiss Res 194:17–35.

Hendrickson AE, Wagoner W, Cowan WM (1972) An autoradiographic and electron microscopic study of retinohypothalamic connections. Z Zellforsch 135:1–36.

Hiroshige T, Honma K-I, Watanabe K (1982a) Ontogeny of the circadian rhythm of plasma corticosterone in blind infantile rats. J Physiol (London) 325:493–506.

Hiroshige T, Honma K-I, Watanabe K (1982b) Possible zeitgebers for external entrainment of the circadian rhythm of plasma corticosterone in blind infantile rats. J Physiol (London) 325:507–519.

Hiroshige T, Honma K-I, Watanabe K (1982c) Prenatal onset and maternal modification of the circadian rhythm of plasma corticosterone in blind infantile rats. J Physiol (London) 325:521–532.

Hoffmann K (1978) Effects of short photoperiod on puberty, growth and moult in the Djungarian hamster (*Phodopus sungorus*). J Reprod Fertil 54:29–35.

Hoffmann K (1984) Photoperiodic reaction in the Djungarian hamster is influenced by previous light history. Biol Reprod 34 (Suppl 1):55.

Honma S, Honma K-I, Shirakawa T, Hiroshige T (1984a) Effects of elimination of maternal circadian rhythms during pregnancy on the development of circadian corticosterone rhythm in blinded infantile rats. Endocrinology 114:44–50.

Honma S, Honma K-I, Shirakawa T, Hiroshige T (1984b) Maternal phase setting of fetal circadian oscillation underlying the plasma corticosterone rhythm in rats. Endocrinology 114:1791–1796.

Horton TH (1984a) Growth and maturation in *Microtus montanus*: effects of photoperiods before and after weaning. Can J Zool 62:1741–1746.

Horton TH (1984b) Growth and reproductive development in *Microtus montanus* is affected by the prenatal photoperiod. Biol Reprod 31:499–504.

Horton TH (1985) Cross-fostering of voles demonstrates in utero effect of photoperiod. Biol Reprod 33:934–939.

Hubel DH, Weisel TN (1970) The period of susceptibility to the physiological effects of unilateral eye closure in kittens. J Physiol (London) 206:419–436.

Ifft JD (1972) An autoradiographic study of the time of final division of neurons in rat hypothalamic nuclei. J Comp Neurol 144:193–204.

Jacques SL, Weaver DR, Reppert SM (1987) Penetration of light into the uterus of pregnant mammals. Photochem Photobiol 45:637–641.

Jolly A (1972) Hour of birth in primates and man. Folia Primat 18:108–121.

Kaiser IH, Halberg F (1962) Circadian periodic aspects of birth. Ann NY Acad Sci 98:1056–1068.

Karsch FJ, Bittman EL, Foster DL, Goodman RL, Legan SJ, Robinson JE (1984) Neuroendocrine basis of seasonal reproduction. Rec Prog Horm Res 40:185–232.

Klein DC (1972) Evidence for the placental transfer of 3H-acetyl-melatonin. Nature (New Biol) 237:117–118.

Koritsanszky S (1981) Fetal and early postnatal cyto- and synaptogenesis in the suprachiasmatic nucleus of the rat hypothalamus. Acta Morphol Acad Sci Hung 29:227–239.

Kreiger DT, Hause L, Krey LC (1977) Suprachiasmatic nuclear lesions do not abolish food-shifted circadian adrenal and temperature rhythmicity. Science 197:398–399.

Lenn NJ, Beebe B, Moore RY (1977) Postnatal development of the suprachiasmatic nucleus in the rat. Cell Tiss Res 178:463–475.

Lewy AJ, Tetsuo M, Markey SP, Goodwin FK, Kopin IJ (1980) Pinealectomy abolishes plasma melatonin in the rat. J Clin Endocr Metab 50:204–207.

Lincoln DW, Porter DG (1976) Timing of the photoperiod and the hour of birth in rats. Nature (London) 260:780–781.

Meijer JH, Rusak B, Harrington ME (1984) Geniculate stimulation phase shifts hamster circadian rhythms. Soc Neurosci Abs 10:502.

Mitchell JA, Yochim JM (1970) Influence of environmental lighting on duration of pregnancy in the rat. Endocrinology 87:472–480.

Moore RY (1983) Organization and function of a central nervous system oscillator: the suprachiasmatic hypothalamic nucleus. Fed Proc 42:2783–2789.

Moore RY, Lenn NJ (1972) A retinohypothalamic projection in the rat. J Comp Neurol 146:1–14.

Negus NC, Berger PJ, Forslund LJ (1977) Reproductive strategy of Microtus montanus. J Mammal 58:347–353.

Persson HE, Stenberg D (1972) Early prenatal development of cortical surface responses to visual stimuli in sheep. Exp Neurol 37:199–208.

Pinter AJ (1968) Effects of diet and light on growth, maturation and adrenal size of Microtus montanus. Am J Physiol 215:461–466.

Pittendrigh CS, Bruce VG (1957) An oscillator model for biological clocks. In: Rudnick D (ed) Rhythmic and Synthetic Processes in Growth. Princeton University Press, Princeton, NJ, pp 75–109.

Porter RH, Etscorn F (1974) Olfactory imprinting resulting from brief exposure in Acomys cahirinus. Nature (London) 250:732–733.

Porter RH, Ruttle K (1975) The responses of one-day-old Acomys cahirinus pups to naturally occurring chemical stimili. Z Tierpsychol 38:154–162.

Porter RH, Cavallaro SA, Moore JD (1980). Developmental parameters of mother-offspring interactions in Acomys cahirinus. Z Tierpsychol 53:153–170.

Porter RH, Matochik JA, Makin JW (1983) Evidence for phenotype matching in spiny mice (Acomys cahirnus). Anim Behav 31:978–984.

Pratt BL (1981) Naturalistic studies of photoperiodism in Syrian and Djungarian hamsters. Ph.D. dissertation, University of Connecticut, Storrs.

Reppert SM, Schwartz WJ (1983) Maternal coordination of a fetal biological clock in utero. Science 220:969–971.

Reppert SM, Schwartz WJ (1984a) Functional activity of the suprachiasmatic nuclei in the fetal primate. Neurosci Lett 46:145–149.

Reppert SM, Schwartz WJ (1984b) The suprachiasmatic nuclei of the fetal rat: characterization of a functional circadian clock using 14C-labeled deoxyglucose. J Neurosci 4:1677–1682.

Reppert SM, Schwartz WJ (1986a) Maternal endocrine extirpations do not abolish maternal coordination of the fetal circadian clock. Endocrinology 119:1763–1767.

Reppert SM, Schwartz WJ (1986b) The maternal suprachiasmatic nuclei are necessary for maternal coordination of the developing circadian system. J Neurosci 6:2724–2729.

Reppert SM, Uhl GR (1987) Vasopressin messenger ribonucleic acid in the supraoptic and suprachiasmatic nuclei: appearance and circadian regulation during development. Endocrinology 120:2483–2487.

Reppert SM, Chez RA, Anderson A, Klein DC (1979) Maternal-fetal transfer of melatonin in a non-human primate. Pediatr Res 13:788–791.

Reppert SM, Coleman RJ, Heath HW, Swedlow JR (1984) Pineal N-acetyltransferase activity in 10-day-old rats: a paradigm for studying the developing circadian system. Endocrinology 115:918–925.

Reppert SM, Duncan MJ, Goldman BD (1985) Photic influences on the developing mammal. In: Photoperiodism, Melatonin and the Pineal (Ciba Foundation Symposium 117), Pitman, London, pp 116–128.

Reppert SM, Henshaw D, Schwartz WJ, Weaver DR (1987a) The circadian-gated timing of birth in rats: disruption by maternal SCN lesions or by removal of the fetal brain. Brain Res 403:398–402.

Reppert SM, Schwartz WJ, Uhl GR (1987b) Arginine vasopressin: a novel peptide rhythm in cerebrospinal fluid, TINS. 10:76–80.

Ribak GE, Peters A (1975) An autoradiographic study of the projections from the lateral geniculate body of the rat. Brain Res 92:341–368.

Rossendale PD, Short RV (1967) The timing of foaling of thoroughbred mares. J Reprod Fertil 13:341–343.

Schwartz WJ, Gainer H (1977) Suprachiasmatic nucleus: use of 14C-labeled deoxyglucose uptake as a functional marker. Science 197:1089–1091.

Schwartz WJ, Davidsen LC, Smith CB (1980) In vivo metabolic activity of a putative circadian oscillator, the rat suprachiasmatic nucleus. J Comp Neurol 189:157–167.

Sedlacek J (1971) Cortical responses to visual stimulation in the developing guinea pig during prenatal and perinatal period. Physiol Bohemoslov 20:213–220.

Sofroniew MV, Weindl A (1980) Identification of parvocellular vasopressin and neurophysin neurons in the suprachiasmatic nucleus of a variety of mammals including primates. J Comp Neurol 193:659–575.

Stanfield B, Cowan WM (1976) Evidence for a change in the retinohypothalamic projection in the rat following early removal of one eye. Brain Res 104:129–136.

Stephan FK (1981) Limits of entrainment to periodic feeding in rats with suprachiasmatic lesions. J Comp Physiol 143:401–410.

Stetson MH, Elliott JA, Goldman BD (1986) Maternal transfer of photoperiodic information influences the photoperiodic response of prepubertal Djungarian hamsters (Phodopus sungorus sungorus). Biol Reprod 34:664–669.

Swanson LW, Cowan WM, Jones EG (1974) An autoradiographic study of the efferent connections of the ventral lateral geniculate nucleus in the albino rat and the cat. J Comp Neurol 156:143–164.

Takahashi JS, Zatz M (1982) Regulation of circadian rhythmicity. Science 217:1104–1111.

Takahashi K, Deguchi T (1983) Entrainment of the circadian rhythms of blinded infant rats by nursing mothers. Physiol Behav 31:373–378.

Takahashi K, Hayafuji C, Murakami N (1982) Foster mother rat entrains circadian adrenocortical rhythm in blinded pups. Am J Physiol 243:E443–E449.

Takahashi K, Murakami N, Hayafuji C, Sasaki Y (1984) Further evidence that circadian rhythm of blinded rat pups is entrained by the nursing dam. Am J Physiol 246:R359–R363.

Tamarkin L, Hollister CW, Lefebvre NG, Goldman B (1977) Melatonin induction of gonadal quiescence in pinealectomized golden hamsters. Science 198:953–955.

Tamarkin L, Reppert SM, Orloff DJ, Klein DC, Yellon SM, Goldman BD (1980). Ontogeny of the pineal melatonin rhythm in the Syrian (*Mesocricetus auratus*) and Siberian (*Phodopus sungorus*) hamsters and in the rat. Endocrinology 107:1061–1064.

Tamarkin L, Baird CJ, Almeida OFX (1985) Melatonin: a coordinating signal for mammalian reproduction? Science 227:714–720.

Terubayashi H, Fujisawa H, Itoi M, Ibata Y (1985) HRP histochemical detection of retinal projections to the hypothalamus in neonatal rats. Acta Histochem Cytochem 18:433–438.

Uhl GR, Reppert SM (1986) Suprachiasmatic nucleus vasopressin messenger RNA: circadian variation in normal and Brattleboro rats. Science 232:390–393.

Viswanathan N, Chandrashekaran MK (1984) Mother mouse sets the circadian clock of pups. Proc Indian Acad Sci 93:235–241.

Viswanathan N, Chandrashekaran MK (1985) Cycles of presence and absence of mother mouse entrain the circadian clock of pups. Nature (London) 317:530–531.

Weaver DR, Reppert SM (1986) Maternal melatonin communicates daylength to the fetus in Djungarian hamsters. Endocrinology 119:2861–2863.

Weaver DR, Reppert SM (1987) Maternal-fetal communication of circadian phase in a precocious rodent, the spiny mouse. Am J Physiol, in press.

Yamazaki J, Takahashi K (1983) Effects of change of mothers and lighting conditions on the development of the circadian adrenocortical rhythm in blinded rat pups. Psychoneuroendocrinology 8:237–244.

Yellon SM, Goldman BD (1984) Photoperiodic control of reproductive development in the male Djungarian hamster (*Phodopus sungorus*). Endocrinology 114:664–670.

Yellon SM, Tamarkin L, Goldman BD (1985) Maturation of the pineal melatonin rhythm in long- and short-day reared Djungarian hamsters. Experientia 41:651–652.

Chapter 9

Neuroendocrine Mechanisms Mediating the Photoperiodic Control of Reproductive Function in Sheep

Robert L. Goodman

I. Introduction

A. Seasonal Breeding in Sheep

The importance of the environment to ovine reproductive function is clearly illustrated by the invariable occurrence of lambing in the spring, when environmental conditions are favorable for the survival of both mother and newborn. In most breeds of sheep, this seasonal pattern in parturition results from annual variations in fertility of the female; ovulatory ovarian cycles occur only in the fall and winter (breeding season) in ewes (Hafez 1952), whereas rams remain fertile all year round. The duration of the ewe breeding season, however, varies considerably among breeds, depending on whether the breed developed under harsh or mild environmental conditions (Robinson 1959). Although in most breeds rams are fertile throughout the year, seasonal variations in reproductive function (e.g., testosterone concentrations) still occur. As is the case for ewes, the magnitude of such annual fluctuations depends on the breed. Indeed, in some primitive breeds (e.g., the Soay) fertility and sexual behavior are limited to a distinct breeding season in the fall (Lincoln and Short 1980).

B. Effects of Photoperiod

The role of environmental factors in controlling seasonal breeding was first conclusively demonstrated by Marshall (1937), who reported that transferring ewes across the equator shifted the timing of breeding and anestrous seasons by six months. Since then, numerous studies have demonstrated that photoperiod is the primary environmental cue controlling reproductive function in sheep. Exposure of ewes to decreasing or short days induces estrous cycles, whereas exposure to increasing or long days suppresses ovarian function (Yeates 1949, Hafez 1952, Mauleon and Rougeot 1962, Thwaites 1965, Ducker and

Bowman 1970, Ducker et al. 1970, Newton and Betts 1972, Legan and Karsch 1980). There is a considerable latency (40–50 days) between a shift in photoperiod and the corresponding reproductive response, but the appropriate response eventually occurs regardless of the time of year. Similarly, short days stimulate testosterone production in the male while long days inhibit testicular function (Pelletier and Ortavant 1975a, Lincoln et al. 1977). In contrast, artificially produced fluctuations in temperature do not alter seasonal reproductive patterns in sheep (Wodzicka-Tomaszewska et al. 1967).

Although it is now generally accepted that photoperiod is the primary environmental cue governing reproductive function in sheep, changes in photoperiod are not actually responsible for alterations in reproductive function at the transitions between breeding and nonbreeding seasons. For example, ewes maintained on a long-day photoperiod equivalent to that of the summer solstice from June 21 through October entered the breeding season at the same time as ewes exposed to natural decreasing daylength during this period (Robinson et al. 1985b). It thus appears that *refractoriness* to long days, rather than stimulation by short days, initiates ovarian cycles at the beginning of the breeding season. An analogous study of the transition to anestrus has led to the conclusion that refractoriness to the stimulatory effects of short days terminates estrous cycles at the end of the breeding season (Robinson and Karsch 1984). A similar photorefractoriness has been recognized for some time in photoperiodic rodents (Reiter 1972). From an adaptive point of view, the development of photorefractoriness allows animals to anticipate the normal shifts in photoperiod; given the latency for the effects of photoperiod, estrous cycles would not begin until mid-December if the animals required a short daylength (i.e., <12 hr) to initiate reproductive function. The Soay ram represents an even more striking example of photorefractoriness (Almeida and Lincoln 1984). To reach peak fertility during the rutting season (October), the testes must start to develop during the summer months. Indeed, clear evidence of increased testicular activity is apparent by July (Lincoln and Short 1980), long before the ram could be stimulated by short daylength.

The photorefractoriness of sheep differs significantly from the more commonly studied photorefractoriness of long-day breeding rodents (Reiter 1972, Stetson and Tate-Ostroff 1981). As already noted, sheep become photorefractory to both long and short daylengths, whereas rodents become refractory only to short days. Moreover, rodents require exposure to long days to be able to respond to short days, whereas sheep do not. Instead, sheep appear to spontaneously "escape" from photorefractoriness without any change in photoperiod (Almeida and Lincoln 1984, Ducker et al. 1973). This ability to escape from photorefractoriness may account for the persistence of cyclic variations in gonadal function in sheep maintained under constant photoperiods (Radford 1961, Thwaites 1965, Wodzicka-Tomaszewska et al. 1967, Ducker et al. 1973, Howles et al. 1982, Almeida and Lincoln 1984). Alternatively, constant photoperiod may allow the expression of an endogenous circannual rhythm in reproductive function (Robinson et al. 1985b). However, it is important to realize that in the absence of photoperiodic cues, these cyclic variations in reproduction are no longer tightly synchronized with the external environment (Legan and Karsch 1983, Bittman et al. 1983a, Almeida and Lincoln 1984). Thus the processing of photoperiodic information is critical for the normal seasonal pattern of ovine reproduction.

II. Processing of Photoperiodic Information in Sheep: An Overview

A. Endocrine Changes Induced by Photoperiod

To analyze the processing of photoperiodic information in sheep, we must first understand the endocrine changes responsible for seasonal variations in gonadal function. In the ewe, the major endocrine change is a dramatic seasonal variation in the negative feedback actions of estradiol on tonic luteinizing hormone (LH) secretion (Legan et al. 1977). This variation in the inhibitory actions of estradiol is illustrated in the top portion of Figure 9-1, which depicts serum LH concentrations throughout the year in two groups of ovariectomized ewes: one group (OVX + E) received Silastic capsules containing estradiol that produced constant estradiol levels (bottom panel), while the other group (OVX) did not.

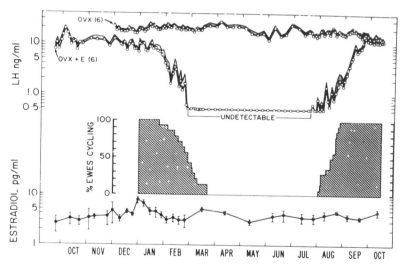

Figure 9-1. Seasonal change in response to estradiol negative feedback. Top panel depicts mean + SEM (shaded area) LH concentrations throughout the year in ovariectomized (OVX) or estradiol-treated ovariectomized (OVX + E) ewes. Estradiol was administered with Silastic capsules that produced relatively constant serum estradiol levels throughout the year (bottom panel). The histogram in the middle panel indicates the breeding and anestrous seasons in a separate group of 14 intact ewes. Redrawn from Legan et al. (1977).

In the estradiol-treated ewes, serum LH concentrations were relatively high from October through January; they then plummeted to undetectable concentrations in the spring and remained suppressed until the fall, when they rose again to concentrations similar to those in untreated ovariectomized ewes. Since LH concentrations in the OVX group did not vary with time, this dramatic seasonal variation in LH concentrations must reflect a corresponding shift in estradiol negative feedback. In the fall and winter, estradiol is a weak negative feedback steroid, whereas in the spring and summer it is a potent inhibitor of tonic LH release. Further, the seasonal shifts in estradiol negative feedback closely coincided with the transitions between breeding and anestrous seasons (Figure 9-1, histogram) and are under photoperiodic control (Legan and Karsch 1980). These observations led to the hypothesis that changes in photoperiod control ovarian function in the ewe by modulating the inhibitory actions of estradiol on tonic LH secretion (Legan et al. 1977). Despite one conflicting observation (Kennaway et al. 1984), the preponderance of experimental evidence provides compelling support for this hypothesis (Karsch et al. 1980).

More recently, it has become evident that photoperiod, in addition to altering estradiol negative feedback, influences LH secretion in untreated ovariectomized ewes (Goodman et al. 1982, Montgomery et al. 1986). These "steroid-independent" actions of photoperiod involve relatively subtle changes in the patterns of pulsatile LH secretion (see below), but they are noteworthy because similar effects have been observed in castrated rams. In the ram, inhibitory photoperiod suppresses both the postcastration rise in LH (Lincoln and Short 1980) and mean LH concentrations in chronically orchidectomized animals (Pelletier and Ortavant 1975b, Lincoln and Short 1980). Inhibitory photoperiod also increases the negative feedback action of testosterone in the ram (Pelletier and Ortavant 1975b). Thus, in both sexes, photoperiod exerts steroid-dependent and steroid-independent actions on LH secretion, but the steroid-independent action is stronger in the ram than in the ewe. Similar steroid-dependent and -independent changes in gonadotropin secretion have been observed in a variety of birds and mammals (Turek and Campbell 1979, Goodman and Karsch 1981a) and, interestingly, the magnitude of the steroid-independent actions of photoperiod varies considerably among species (Goodman and Karsch 1981a), as it does between ewes and rams. The endocrine actions of photoperiod in ewes (Karsch 1980) and rams (Lincoln and Short 1980) have already been extensively reviewed, so we will focus on the neuroendocrine mechanisms by which photoperiod controls these changes in gonadoptropin secretion.

B. Major Steps in the Processing of Photoperiodic Information

The neuroendocrine mechanisms by which a change in daylength in the external environment induces changes in the control of gonadotropin secretion can be divided into four major steps. The first step is *perception* of the length of the day, a process that requires both a photoreceptor and a mechanism for measurement of daylength. *Transmission* of photoperiodic information from the photoreceptor to the pineal gland then takes place via a neural pathway. The third step in this process is *transduction* of this neural information into an endocrine signal by the pineal gland. Finally, *translation* of the pineal endocrine signal into a change in gonadotropin secretion by the hypothalamo-hypophyseal axis completes this process. It is this final step that is manifest as the steroid-dependent and steroid-independent effects of photoperiod described above. Although there is now general agreement that these steps comprise the processing of photoperiodic information in sheep, some steps have been much better characterized than others. In the rest of this review, we will consider the evidence for each step and, when possible, describe in detail the underlying physiological mechanisms.

III. Perception of Daylength

A. The Photoreceptor

No information is available on the location of the photoreceptor in rams, but one study in ewes has provided strong evidence that retinal photoreceptors are critical for seasonal breeding (Legan and Karsch 1983). In this study, ewes were blinded by bilateral enucleation, placed into photoperiod-control rooms, and subjected to 90-day alterations between long (16L:8D) and short (8L:16D) days. The effect of these photoperiod manipulations on estradiol negative feedback is illustrated in Figure 9-2, which presents serum LH concentrations in estradiol-treated ovariectomized ewes during an 18-month period of this study. In control ewes, each photoperiod shift produced the appropriate response; LH concentrations increased (low estradiol negative feedback) in short days and decreased in long days (high estradiol negative feedback). Blinded ewes, in contrast, were no longer able to respond to photoperiod (Figure 9-2, middle panel). However, LH concentrations did not become acyclic in the absence of photoperiod cues. Instead, throughout most of this study, LH levels showed an annual rhythm, with elevated LH levels in fall and winter and low LH concentrations in spring and summer. Analogous results were observed in ovary-intact blinded ewes; there was a circannual variation in the occurrence of estrous cycles that was not influenced by photoperiod (Legan and Karsch 1983).

From these observations it seems reasonable to conclude that in the sheep, as in other mammals (Turek and Campbell 1979), the eyes are the photoreceptors essential for entrainment of reproductive function by photoperiod. Reproductive function, however, can still be entrained by external factors in blinded ewes. For example, cues from sighted rams, probably pheromones (Martin and Scaramuzzi 1983), will synchronize ovarian function in blinded sheep (Legan and Karsch 1983) and may account for the persistence of seasonal breeding in blinded ewes maintained outdoors with sighted rams (Clegg et al. 1965). This raises the possibility that the circannual variation in reproductive function observed in the absence of photoperiodic and pheromonal cues (Figure 9-2) reflects entrainment by another environmental factor (e.g., temperature). However, because this circannual variation tends to become asynchronous with the envi-

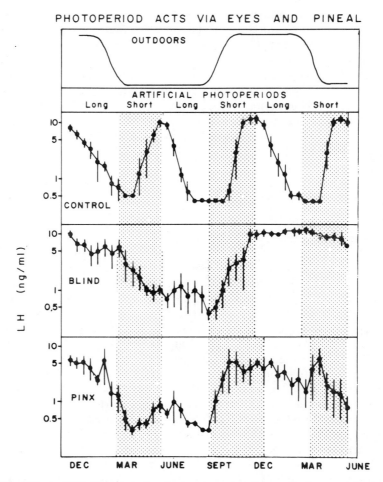

Figure 9-2. Effect of blinding and pinealectomy on the photoperiodic control of estradiol negative feedback. Bottom three panels present mean (± SEM) LH concentrations in three groups of estradiol-treated ovariectomized ewes: CONTROL (sighted, pineal-intact), BLIND (orbital enucleation), or PINX (pinealectomized). Animals were maintained in photoperiod control rooms and subjected to alternating 90-day periods of long (16L:8D) and short (8L:16D, shading) days, except for one 120-day period (Sept.–Dec.) of short-day treatment of PINX ewes. LH concentrations in estradiol-treated ovariectomized ewes maintained outdoors are depicted in the top panel for comparison purposes. Data from control and blinded ewes redrawn from Legan and Karsch (1983); PINX data from Bittman et al. (1983a).

ronment (Legan and Karsch 1983), it most likely reflects an endogenous circannual rhythm. Interestingly, similar circannual variations in gonadal function are observed in sheep exposed to constant photoperiod for prolonged periods (Radford 1961, Ducker et al. 1973, Howles et al. 1982, Almeida and Lincoln 1984). Thus, perception of photoperiod information by sighted sheep may require a *change* in daylength.

B. Photoperiodic Time Measurement

Perception of daylength requires both a photoreceptor and a method for measuring the length of the day. In a large number of birds, mammals, plants, and insects, daylength is measured using an endogenous circadian rhythm in sensitivity to light, a concept now known as the "Bünning hypothesis" (Bünning 1936, Follett and Follett 1981). In its simplest form, this endogenous rhythm is entrained by the 24-hour light–dark cycle so that the period of photosensitivity begins more than 12 hours after the onset of light. If light occurs during the photosensitive period, the day is perceived as long (i.e., greater than 12 hours of light); if no light occurs, then the day is perceived as short (Turek and Campbell 1979).

To date, there have only been a few experiments that have tested this hypothesis in sheep, but the results obtained have all been consistent with it. Perhaps the strongest evidence for the existance of a circadian rhythm in photosensitivity in sheep comes from a "resonance-type" experiment performed by Almeida and Lincoln (1982). They observed that shifting rams from a 16L:8D to an 8L:40D photoperiod induced testicular growth, just as a shift to an 8L:16D photoperiod did. In contrast, a photoperiod shift from 16L:8D to 8L:28D had no stimulatory effects. These results demonstrate that a reduction in the amount of light/day need not be interpreted as a shift from long to short days and are consistent with the use of an endogenous circadian rhythm in photosensitivity. Light pulse experiments, in which the dark phase is interrupted at different times by a brief period of illumination, have also produced results largely consistent with this hypothesis (Garnier et al. 1977, Lincoln 1978, Thimonier et al. 1985). It thus appears likely that sheep measure daylength using an endogenous circadian rhythm in photosensitivity, although further work is clearly needed to characterize this rhythm.

IV. Transmission of Photoperiodic Information to the Pineal Gland

The existence of a pathway to transmit information from the retinal photoreceptors to the pineal gland in sheep has been inferred largely from studies on the role of the pineal in mediating the effects of photoperiod on reproductive function. Several studies have demonstrated that pinealectomy blocks the ability of photoperiod to influence reproductive function in both rams (Barrell and Lapwood 1979a,b) and ewes (Bittman et al. 1983a,b, Bittman and Karsch 1984, Yellon et al. 1985). The results of one such study (Bittman et al. 1983a) are illustrated in Figure 9-2 (bottom panel), which depicts the effects of pinealectomy on photoperiod-induced shifts in estradiol negative feedback. Note that removal of the pineal had exactly the same effect as removal of the retinal photoreceptors; LH concentrations in these estradiol-treated ovariectomized ewes were no longer influenced by photoperiod but still showed a circannual variation which tended to become asynchronous with time after pinealectomy. Thus pinealectomy, like blinding, does not abolish seasonal fluctuations in reproductive function (Roche et al. 1970, Kennaway et al. 1981). It does, however, prevent the entraining effects of photoperiod, and appears to do so, at least in ewes, by altering estradiol negative feedback (Figure 9-2). This concept has recently been challenged because chronically pinealectomized ewes showed normal seasonal reproductive patterns but no obvious seasonal shifts in estradiol negative feedback (Kennaway et al. 1984). It should be noted, however, that these ewes were maintained with pineal-intact rams so that the persistence of seasonal reproductive patterns may well have been due to pheromonal cues (Legan and Karsch 1983). This study thus points out the need for further work on the nature and mechanism of action of such pheromonal cues, but does not rule out the importance of changes in estradiol negative feedback to the photoperiodic control of reproduction.

The similarity in the effects of blinding and pinealectomy, together with the role of photoperiod in controlling pineal gland function (see below), implies that information is transmitted from the retina to the pineal (Lincoln and Short 1980, Legan and Winans 1981, Karsch et al. 1984). However, the exact anatomical route by which neural impulses from the retina reach the pineal in sheep remains largely unknown. In the hamster, this pathway includes the suprachiasmatic nucleus (SCN), the paraventricular nucleus, and the superior cervical ganglion (SCG), which innervates the pineal (Stetson and Watson-Whitmyre 1976, Reiter 1980, Turek et al. 1984). In the sheep, there is strong evidence that this pathway includes the SCG; removal of this ganglion blocks the effects of photoperiod on both reproductive (Lincoln 1979, Lincoln and Almeida 1982) and pineal (Lincoln et al. 1982) function in the ram. The SCN may also be involved since this nucleus receives a direct neural projection from the retina (Legan and Winans 1981) and SCN lesions disrupt seasonal breeding patterns in ewes (Przekop 1978, Przekop and Domanski 1980, Pau et al. 1982). However, no study has examined the effect of SCN lesions on pineal function in sheep, so that a role for this nucleus in transmitting photoperiodic information to the pineal remains to be conclusively demonstrated.

V. Transduction of Photoperiodic Information by the Pineal Gland

Based on the effects of pinealectomy (e.g., Figure 9-2), it is now generally accepted that this gland plays a crucial role in mediating the effects of photoperiod (Turek and Campbell 1979, Karsch et al. 1984). Since the pineal gland has no neural efferents (Ariens-Kappers 1976), it must exert its effects on reproductive function hormonally. Because of the large amount of work implicating melatonin as an important pineal hormone in other species (Reiter 1980), most of the work on the role of the pineal in sheep has focused on this indoleamine. In general, this work falls into two general categories: the control of melatonin secretion by photoperiod and the effects of melatonin on reproductive function.

A. Control of Melatonin Secretion

The secretion of melatonin by the pineal gland represents a circadian rhythm that is tightly controlled by the photoperiod. The circadian nature of melatonin secretion has been clearly demonstrated by the persistence of a 24-hour rhythm in serum melatonin concentrations following transfer of ewes to constant darkness (Rollag and Niswender 1976, Earl et al. 1985). Photoperiod has two effects on this circadian rhythm. First, it synchronizes the rhythm to the 24-hour light–dark cycle (Rollag and Niswender 1976, Kennaway et al. 1982a, Almeida and Lincoln 1982, Bittman et al. 1983b). Second, light directly inhibits melatonin release (Rollag et al. 1978, Earl et al. 1985) so that melatonin levels remain low as long as light remains on (Rollag and Niswender 1976). The overall effect of these two actions of photoperiod is that melatonin concentrations increase shortly after the onset of darkness and remain elevated until the beginning of the light phase (Figure 9-3, top panel). This melatonin pattern has now been observed in a large number of studies using both natural and artificial photoperiods (Lincoln et al. 1982, Bittman et al. 1983b, Bittman and Karsch 1984, Rollag and Niswender 1976, Rollag et al. 1978, Kennaway et al. 1977, Arendt et al. 1981, Kennaway et al. 1982b, 1983). Thus, the serum melatonin pattern can be considered an internal hormonal analogue of the light–dark cycle in the external environment.

B. Role of Melatonin in Control of Reproductive Function

Several different approaches have been used to assess the importance of melatonin to the photoperiodic control of reproductive function, and all of them have provided support for the proposal that the melatonin pattern mediates the effects of photoperiod in sheep. First, as already discussed, abolition of the diurnal variation in serum melatonin by removal of either the SCG (Lincoln 1979, Lincoln et al. 1982) or the pineal (Kennaway et al. 1977, Barrell and Lapwood 1979a,b, Bittman et al. 1983a,b) disrupts the effects of photoperiod. Second, changes in photoperiod that alter reproductive function also alter melatonin patterns, even when abormal photoperiods are used (Lincoln et al. 1982, Almeida and Lincoln 1982, Kennaway et al. 1983). For example, exposure of rams to an 8L:40D photoperiod produced a melatonin pattern similar to that seen with an 8L:16D photoperiod and also had the same stimulatory effects on testicular function (Almeida and Lincoln 1982).

The third approach that has been used is melatonin administration. In a number of studies, melatonin has been administered to pineal-intact animals at a time of day that prolongs the endogenous melatonin peak (Kennaway et al. 1982c, Nett and Niswender 1982, Arendt et al. 1983, Lincoln and Ebling 1985). When this is done to animals kept on long days, it creates a short-day-type melatonin pattern, which should stimulate reproductive function if the melatonin pattern is the critical pineal signal. In fact, such melato-

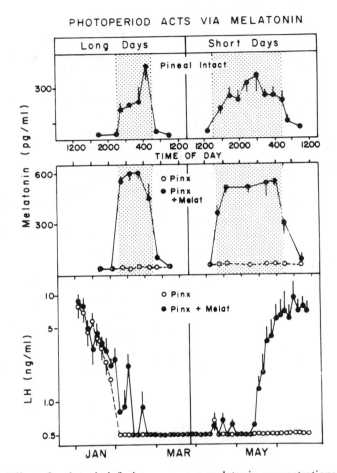

Figure 9-3. Effect of melatonin infusions on serum melatonin concentrations and estradiol negative feedback in ewes. Top panel: Mean (±SEM) melatonin concentrations throughout the day in pineal-intact ewes maintained on either long (16L:8D) or short (8L:16D) days; shading indicates the period of darkness. Middle panel: Serum melatonin levels throughout the day in pinealectomized (Pinx) ewes that were maintained on long (left) and short (right) days and received either melatonin infusions during the dark (filled circles) or were not infused with melatonin (open circles). Bottom panel: Mean (±SEM) LH concentrations in Pinx estradiol-treated ovariectomized ewes receiving either nightly melatonin infusions (filled circles) or no melatonin treatment (open circles). Note that the time scale is months, not hours. During the first 3 months, ewes were maintained on long days with melatonin administered during the 8 hours of darkness (middle panel). On March 21 (vertical line), photoperiod was switched to short days and the melatonin infusion prolonged to cover the 16 hours of darkness (middle panel). Data replotted from Bittman et al. (1983b).

nin treatment advances the onset of reproductive function in both rams (Lincoln and Ebling 1985) and ewes (Kennaway et al. 1982c, Nett and Niswender 1982, Arendt et al. 1983). While these data provide support for the hypothesis that melatonin controls reproductive function, they do not preclude a role for other pineal hormones since pineal-intact sheep were used. Further, the role of melatonin in mediating the effects of long days cannot be assessed by this approach, because a long-day melatonin pattern cannot be produced during short days in pineal-intact animals. To overcome these limitations, Bittman and coworkers examined the effects of nightly melatonin infusions in pinealectomized ewes (Bittman et al. 1983b, Bittman and Karsch 1984, Yellon et al. 1985). As illustrated in Figure 9-3, melatonin infusions (middle panel) produced a reasonably good approximation of the normal melatonin patterns in both long and short days (top panel). Further, shifting the melatonin pattern was able to mimic the effects of a shift in photoperiod in pineal-intact ewes. Specifically, LH concentrations in estradiol-treated ovariectomized ewes increased in response to a short-day melatonin pattern (Figure 9-3, bottom panel) just as they do in response to a short photoperiod in pineal-intact ewes (e.g., Figure 9-2). Subsequent studies have demonstrated that long-day melatonin patterns inhibit LH secretion in estradiol-treated ovariectomized ewes (Bittman and Karsch 1984). Moreover, these melatonin patterns are effective regardless of the ambient photoperiod (Bittman and Karsch 1984, Yellon et al. 1985). Thus, the secretion of melatonin appears to be sufficient to account for the role of the pineal in mediating the effects of photoperiod in the ewe.

Although there is now compelling evidence that the diurnal melatonin pattern is the critical endocrine signal produced by the pineal gland, the characteristic of this rhythm that encodes for daylength remains to be determined. Three characteristics of the melatonin pattern have been suggested: the amplitude (Lincoln et al. 1981), the duration (Bittman and Karsch 1984), and the phase relative to the 24-hour light–dark cycle (Almeida and Lincoln 1982). Of these three, amplitude seems the least likely, since the amplitude of the melatonin pattern does not vary significantly between long and short days (Rollag and Niswander 1976, Rollag et al. 1978, Kennaway et al. 1983, Lincoln et al. 1982, Bittman et al. 1983b). In contrast, the duration of melatonin secretion clearly varies directly with the length of the night (Figure 9-3), and the results of melatonin administration experiments have all been consistent with duration encoding for daylength (e.g., Bittman and Karsch 1984, Yellon et al. 1985). These studies, however, have not ruled out a role for the phase of the melatonin rise, and there are some data suggesting that this characteristic may be important (Almeida and Lincoln 1982). Regardless of whether the phase or duration of the melatonin rise is crucial, it is important to realize that melatonin is neither pro- nor antigonadal in the sheep. Instead, melatonin secretion mediates the effects of both stimulatory and inhibitory photoperiod so that the melatonin pattern represents a hormonal code for the 24-hour light–dark cycle in the external environment.

VI. Translation of the Melatonin Pattern into a Change in Gonadotropin Secretion by the Hypothalamo-Hypophyseal Axis

At this time, there is virtually no information available on the sites of action and mechanisms by which melatonin controls gonadotropin secretion in ewes or rams. However, we have recently developed a model for the seasonal alterations in hypothalamo-hypophyseal function that underlie seasonal breeding in the ewe. Since these changes are presumably mediated by melatonin, this model may provide some insight into the actions of this indoleamine in sheep. This model grew out of two observations on the control of the pulsatile LH release that comprises tonic LH secretion in the ewe (Butler et al. 1972).

A. Role of LH Pulses in Control of Seasonal Breeding

The first observation was that both the steroid-independent and steroid-dependent actions of photoperiod reflect changes in the control of LH pulse frequency. In untreated ovariectomized ewes there is a slight, but significant, decrease in LH pulse frequency in anestrus (Figure 9-4, top panel) relative to that seen during the breeding season (Goodman et al. 1982, Montgomery et al. 1985). There is a compensatory increase in LH pulse amplitude so that mean LH concentrations are not lower (Figure 9-1), and in some cases may actually be higher (Pau and Jackson 1985), during inhibitory photoperiods. This seasonal shift in pulse frequency is observed in some breeds of sheep (Goodman et al. 1982, Montgomery et al. 1985) but not in Merino ewes (Martin et al. 1983), which have a relatively brief anestrous period. LH pulse frequency in ovariectomized ewes is controlled by photoperiod; long days decrease and short days increase pulse frequency (Robinson et al. 1985a). Moreover, these effects of photoperiod appear to be mediated by the melatonin pattern (Bittman et al. 1982). Thus, the anestrous-associated decrease in LH pulse frequency has all the characteristics expected of the steroid-independent action of photoperiod.

The seasonal variation in estradiol negative feedback is also due to a corresponding shift in the ability of this steroid to suppress LH pulse frequency (Figure 9-4, middle panel). During the breeding season, when estradiol is a weak inhibitory steroid, it suppresses LH pulse amplitude but cannot suppress pulse frequency (Goodman and Karsch 1980, Goodman et al. 1982, Martin et al. 1983, Wright et al. 1981). In anestrus, however, estradiol gains the capacity to suppress LH pulse frequency, and consequently becomes a potent negative feedback steroid (Goodman et al. 1982, Martin et al. 1983). Thus, the seasonal variation in the inhibitory effects of estradiol depicted in Figure 9-1 reflects a qualitative shift in the negative feedback action of this steroid. In contrast to estradiol, progesterone can inhibit LH pulse frequency during both the breeding (Figure 9-4) and anestrous seasons (Goodman and Karsch 1980, Goodman et al. 1982). The actions of this steroid in anestrus, however, are of no physiological significance since circulating progesterone concentrations are undetectable at this time of year. Finally, it should be noted that since changes in LH pulse frequency are most likely due to changes in the frequency of gonadotropin-

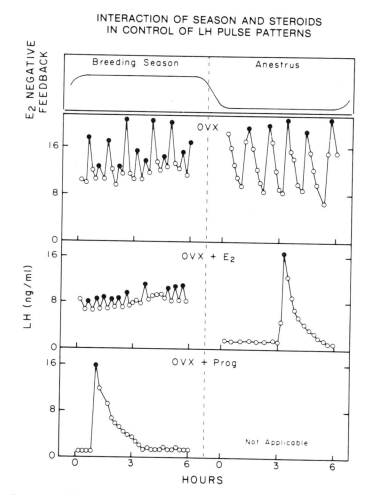

Figure 9-4. Seasonal differences in episodic LH secretion in untreated (top panel), estradiol (E_2)-treated (middle panel), and progesterone (Prog)-treated (bottom panel) ovariectomized (OVX) ewes. Filled circles depict peaks of statistically identified LH pulses. Seasonal change in estradiol negative feedback is illustrated at top for comparison purposes. Data in bottom three panels are representative pulse patterns taken from Goodman and Karsch (1980) and Goodman et al. (1982).

releasing hormone (GnRH) pulses (Levine et al. 1982, Clarke and Cummings 1982), both steroid-independent and steroid-dependent actions of photoperiod reflect alterations in hypothalamic function.

B. Model for the Seasonal Control of Pulsatile LH Secretion

The second observation that led to our model came from studies on the effects of pentobarbital anesthesia in the ewe (Goodman and Meyer 1984). Pentobarbital

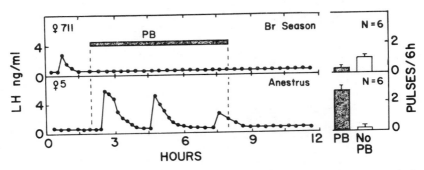

Figure 9-5. Effect of pentobarbital (PB) anesthesia on pulsatile LH secretion in intact anestrous (bottom panel) and luteal-phase (top panel) ewes. Left portion depicts LH concentrations from representative ewes before, during, and after pentobarbital administration (horizontal bar). Bars on right represent mean LH pulse frequency during anesthesia (shaded bar) or during a corresponding period in unanesthetized control ewes. Redrawn from Goodman and Meyer (1984).

administration to ovary-intact anestrous ewes produced an immediate and dramatic increase in tonic LH secretion by specifically increasing LH pulse frequency (Figure 9-5). In contrast, pentobarbital anesthesia of luteal-phase ewes during the breeding season suppressed episodic LH release. Pentobarbital had similar, but not as dramatic, effects in ovariectomized ewes; it increased LH pulse frequency slightly in anestrus while decreasing it during the breeding season (Goodman and Meyer 1984). In interpreting these data, we assumed that pentobarbital acts by decreasing neuronal activity (Barker and Ranson 1978). If this assumption is correct, then one can conclude that during anestrus LH pulse frequency is held in check by a set of inhibitory neurons; pentobarbital decreases the activity of these inhibitory neurons, allowing pulsatile LH secretion to increase. Further, since pentobarbital did not increase LH in the breeding season, these inhibitory neurons may be inactive at this time of year.

The inference drawn from the effects of pentobarbital led to the following model to account for the seasonal variations in control of pulsatile LH secretion (Figure 9-6). The key postulate of this model is that the long-day photoperiod of anestrus activates a neural system that inhibits the frequency of the hypothalamic "pulse generator" controlling GnRH release in the ewe (Goodman and Karsch 1981b, Lincoln et al. 1985). In addition, this model proposes that these anestrous inhibitory neurons can be stimulated by estradiol (Goodman and Meyer 1984). Estradiol thus suppresses LH pulse frequency by increasing the activity of these neurons, but can only do so in anestrus. In addition, the low activity of these inhibitory neurons in the absence of estradiol produces the slight decrease in LH pulse frequency seen in ovariectomized ewes during anestrus. Thus, this model can account for both the steroid-independent and steroid-dependent actions of photoperiod.

As a first test of this model, we determined what neurotransmitters are released by these putative inhibitory neurons. This was done by screening eight

ANESTRUS   BREEDING SEASON

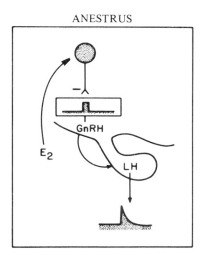   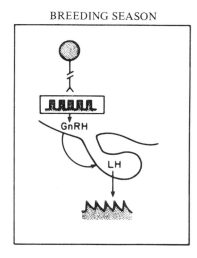

Figure 9-6. Model for the seasonal control of tonic LH secretion in the ewe. During anestrus (left panel), long-day photoperiod activates a set of estradiol-sensitive neurons that inhibit the activity of the GnRH pulse generator (box). Thus, estradiol can inhibit pulse frequency in anestrus, but it cannot do so during the breeding season when these inhibitory neurons are inactive (right panel). Modified from Goodman and Meyer (1984).

neurotransmitter antagonists for their ability to increase pulsatile LH secretion in intact anestrous ewes (Meyer and Goodman 1985). Of the eight antagonists tested, only the dopaminergic (pimozide) and the α-adrenergic (phenoxybenzamine) blockers increased LH pulse frequency (Figure 9-7). We next performed a series of experiments, the results of which were consistent with the hypothesis that the steroid-dependent effects of anestrous photoperiod reflect the activation of estradiol-sensitive catecholaminergic neurons (Meyer and Goodman 1985, 1986). This supporting evidence includes the observations that: (1) pimozide and phenoxybenzamine do not increase LH release during the breeding season; (2) other dopaminergic and α-adrenergic antagonists increase LH pulse frequency in intact anestrous ewes; and (3) dopaminergic and α-adrenergic agonists suppress LH release in ovariectomized ewes. Finally, recent data raise the possibility that these two neural systems may be linked "in series" with the α-adrenergic system activating inhibitory dopaminergic neurons (Goodman 1986).

These catecholaminergic systems, however, do not appear to account for the steroid-independent effects of photoperiod. If they did, then pimozide and/ or phenoxybenzamine should increase LH pulse frequency in ovariectomized anestrous ewes. But no such stimulatory actions were observed when several doses of these antagonists were tested in ovariectomized animals (Meyer and Goodman 1985, 1986). Tests of other neurotransmitter antagonists revealed that only cyproheptadine, a serotonin antagonist, increased LH pulse frequency in ovariectomized ewes (Meyer and Goodman 1986, Schillo et al. 1985). Further,

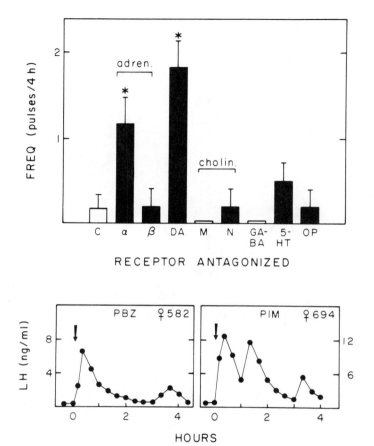

Figure 9-7. Effect of receptor antagonists on LH pulses in intact anestrous ewes. Top panel depicts mean (+ SEM) LH pulse frequency during the 4 hours following injection of eight different neurotransmitter receptor antagonists. Control (C) ewes were not injected. The following antagonists were used to block specific receptors: α-adrenergic (α)—phenoxybenzamine; β-adrenergic (β)—propranolol; dopaminergic (DA)—pimozide; muscarinic cholinergic (M)—atropine; nicotinic cholinergic (N)—mecamylamine; GABAnergic (GABA)—bicuculline; serotonergic (5-HT)—cyproheptadine; and opiate (OP)—naltrexone. Bottom panel depicts representative LH pulse patterns following injection of phenoxybenzamine (PBZ) and pimozide (PIM). Redrawn from Meyer and Goodman (1985).

this stimulatory effect was observed during anestrus but not the breeding season (Figure 9-8). These data suggest that a serotonergic neural system may mediate the steroid-independent actions of photoperiod. They also raise the possibility that distinct neural systems mediate the steroid-dependent and steroid-independent effects of photoperiod in the ewe. This possibility is supported by two other observations. First, the temporal characteristics of these two actions of

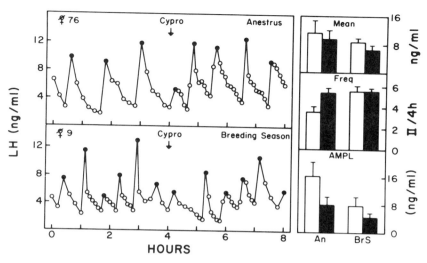

Figure 9-8. Effect of a serotonin antagonist on pulsatile LH secretion in untreated ovariectomized ewes. Left panel presents LH pulse patterns before and after the iv injection of cyproheptadine (Cypro) during anestrus (top) and breeding season (bottom). Bars on the right depict mean (+ SEM) LH pulse amplitude (AMPL), frequency (Freq), and mean LH concentrations (Mean) before (open bars) and after (shaded bars) cyproheptadine injection in anestrus (AN) and breeding season (BrS) ewes. Reprinted from Meyer and Goodman (1986).

photoperiod are quite different; the steroid-independent changes in pulse frequency occur gradually over several months (Robinson et al. 1985a), whereas changes in estradiol negative feedback are abrupt, occurring over a period of two weeks (Legan et al. 1977). Second, knife cuts just posterior to the SCN disrupt the steroid-dependent, but not the steroid-independent, effects of photoperiod (Pau and Jackson 1985). Thus it appears that anestrous photoperiod, acting via melatonin, activates two neuronal systems, both of which inhibit the activity of the hypothalamic GnRH pulse generator: a set of estradiol-sensitive catecholaminergic neurons and a set of serotonergic neurons, which account for the steroid-dependent and steroid-independent suppression of LH, respectively.

This model is also consistent with other available data on the seasonal control of reproductive function in sheep. The concept of an active neural suppression of reproductive function in anestrus is supported by the ability of neural lesions to induce ovulatory cycles during the anestrous season (Przekop 1978, Przekop and Domanski 1980, Pau et al. 1982). Moreover, in the ram, inhibitory photoperiod suppresses reproductive function primarily by decreasing GnRH pulse frequency (Lincoln and Short 1980) and may do so by activating an inhibitory neural system, since anesthesia appears to increase LH in intact rams exposed to long days (Lincoln and Almeida 1982).

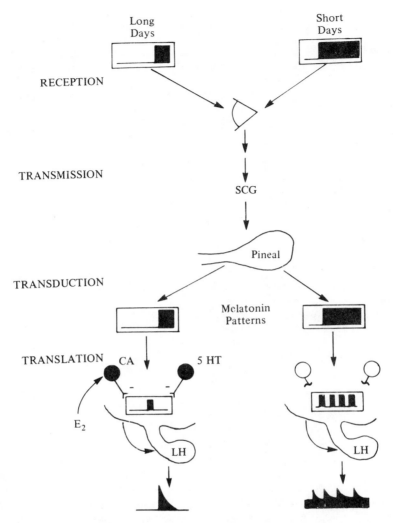

Figure 9-9. Model for the photoperiod control of ovine reproduction.

VII. Conclusion

By drawing on recent work in both rams and ewes a reasonably coherent model can be developed for the mechanisms by which photoperiod controls reproductive function in sheep (Figure 9-9). The first step in the processing of photoperiodic information is reception by retinal photoreceptors. Information is then transmitted to the pineal gland via a pathway that includes the SCG. At the pineal, photoperiodic information is transduced into a hormonal signal, the daily melatonin pattern. The melatonin pattern can be considered a hormonal analogue of the 24-hour light–dark cycle and encodes for both inhibitory long days and stimulatory short days. Finally, the long-day melatonin pattern activates two inhibitory neural systems: a set of estradiol-sensitive catecholaminergic neurons and a set of estradiol-insensitive serotonergic neurons. As a result, the long-day melatonin pattern is translated into a steroid-dependent and steroid-independent alteration in LH pulse frequency.

There are obviously some important gaps in this overall picture. The pathway for transmission of information from the photoreceptor to the pineal and the sites and mechanism of action of melatonin remain largely unknown. Moreover, the mechanisms underlying photorefractoriness and the existence, and significance, of an endogenous circannual rhythm in reproductive function are just beginning to be examined. Finally, it is important to realize that some aspects of this model should be considered working hypotheses that require further experimental testing. For example, a more systematic examination of the effects of different melatonin patterns is needed to determine if the duration of the melatonin rise is the critical characteristic encoding daylength. Similarly, more information on the location and function of the putative inhibitory neural systems is needed before their role in mediating the steroid-dependent and steroid-independent actions of photoperiod can be accepted. Nonetheless, the model depicted in Figure 9-9 provides a useful framework for future studies on the processing of photoperiodic information critical to the control of ovine reproductive function.

Acknowledgments. I thank the following colleagues at the University of Michigan and West Virginia University, whose help and advice has been instrumental in the planning and execution of our work on the seasonal control of pulsatile LH secretion in ewes: Drs. Fred Karsch, Steve Meyer, Sandy Legan, Kathy Ryan, Doug Foster, Keith Inskeep, and Bob Dailey. Studies on the control of LH pulses were supported by a grant from NIH (HD17864).

References

Almeida OFX, Lincoln GA (1982) Photoperiodic regulation of reproductive activity in the ram: evidence for the involvement of circadian rhythms in melatonin and prolactin secretions. Biol Reprod 27:1062–1075.

Almeida OFX, Lincoln GA (1984) Reproductive photorefractoriness in rams and accompanying changes in the patterns of melatonin and prolactin secretion. Biol Reprod 30:143–158.

Arendt J, Symons AM, Laud C (1981) Pineal function in sheep: evidence for a possible mechanism mediating seasonal reproductive activity. Experientia 37:584–586.

Arendt J, Symons AM, Laud CA, Pryde SJ (1983) Melatonin can induce early onset of the breeding season in ewes. J Endocrin 97:395–400.

Ariens-Kappers J (1976) The mammalian pineal gland, a survey. Acta Neurochir 34:109–149.

Barker JL, Ranson BR (1978) Pentobarbitone pharmacology of mammalian central neurons grown in tissue culture. J Physiol 280:355–372.

Barrell GK, Lapwood KR (1979a) Effects of various lighting regimens and pinealectomy on semen production in Romney rams. J Reprod Fert 57:273–279.

Barrell GK, Lapwood KR (1979b) Effects of pinealectomy on the secretion of luteinizing hormone, testosterone and prolactin in rams exposed to various lighting regimes. J Endocrinol 80:397–405.

Bittman EL, Karsch FJ (1984) Nightly duration of pineal melatonin secretion determines the reproductive response to inhibitory day length in the ewe. Biol Reprod 30:585–593.

Bittman EL, Karsch FJ, Dempsey RJ (1982) Pineal melatonin mediates photoperiodic drive of seasonal changes in pulsatile LH secretion in sheep. 64th Annual Meeting, The Endocrine Society, Abstr. 983.

Bittman EL, Karsch FJ, Hopkins JW (1983a) Role of the pineal gland in ovine photoperiodism: regulation of seasonal breeding and negative feedback effects of estradiol upon luteinizing hormone secretion. Endocrinology 113:329–336.

Bittman EL, Dempsey RJ, Karsch FJ (1983b) Pineal melatonin secretion drives the reproductive response to daylength in the ewe. Endocrinology 113:2276–2283.

Bunning E (1936) Die endogene Tagesrhythmik als Grundlage der photoperischen Reaktion. Ber Dtsch Bot Ges 54:590–607.

Butler WR, Malven PV, Willett LB, Bolt DJ (1972) Patterns of pituitary release and cranial output of LH and prolactin in ovariectomized ewes. Endocrinology. 91:793–801.

Clarke IJ, Cummins JT (1982) The temporal relationship between gonadotropin releasing hormone (GnRH) and luteinizing hormone (LH) secretion in ovariectomized ewes. Endocrinology 111:1737–1739.

Clegg MT, Cole HH, Ganong WF (1965) The role of light in the regulation of cyclic estrous activity in sheep. US Dept Agric Misc Pub 1005:96–103.

Ducker MJ, Bowman JC (1970) Photoperiodism in the ewe. 3. The effects of various patterns of increasing daylength on the onset of anoestrus in Clun Forest ewes. Anim Prod 112:467–471.

Ducker MJ, Thwaites CJ, Bowman JC (1970) Photoperiodism in the ewe. 2. The effects of various patterns of decreasing day length on the onset of oestrus in Clun Forest ewes. Anim Prod 12:115–123.

Ducker MJ, Bowman JC, Temple A (1973) The effect of constant photoperiod on the expression of oestrus in the ewe. J Reprod Fert Suppl 9:143–150.

Earl CR, D'Occhio MJ, Kennaway DJ, Seamark RF (1985) Serum melatonin profiles and endocrine responses of ewes exposed to a pulse of light late in the dark phase. Endocrinology 117:226–230.

Follett BK, Follett DE (1981) Biological Clocks in Seasonal Reproductive Cycles. John Wright and Sons, Bristol, England.

Garnier D-H, Ortavant R, Mansard F-X, Terqui M (1977) Influence de la lumière sur les variations de la testostéronemie chez le bélier: mise en evidence d'une phase photosensible au cours du rhythme diurne. C R Acad Sci Ser D 284:61–64.

Goodman RL (1986) Functional organization of the adrenergic and dopaminergic systems inhibiting LH secretion in anestrous ewes. Biol Reprod 34 (Suppl 1):141.

Goodman RL, Karsch FJ (1980) Pulsatile secretion of luteinizing hormone: differential suppression by ovarian steroids. Endocrinology 107:1286–1290.

Goodman RL, Karsch FJ (1981a) A critique of the evidence on the importance of steroid feedback to seasonal changes in gonadotrophin secretion. J Reprod Fert Suppl 30:1–13.

Goodman RL, Karsch FJ (1981b) The hypothalamic pulse generator: a key determinant of reproductive cycles in sheep. In: Follett BK, Follett DE (eds) Biological Clocks in Seasonal Reproductive Cycles. John Wright and Sons, Bristol, England, pp. 221–236.

Goodman RL, Meyer SL (1984) Effects of pentobarbital anesthesia on tonic luteinizing hormone secretion in the ewe: evidence for active inhibition of luteinizing hormone in anestrus. Biol Reprod 30:374–381.

Goodman RL, Bittman EL, Foster DL, Karsch FJ (1982) Alterations in the control of luteinizing hormone pulse frequency underlie the seasonal variation in estradiol negative feedback in the ewe. Biol Reprod 27:580–589.

Hafez ESE (1952) Studies on the breeding season and reproduction of the ewe. J Agric Sci 42:189–265.

Howles CM, Craigon J, Haynes NB (1982) Long-term rhythms of testicular volume and plasma prolactin concentrations in rams reared for 3 years in constant photoperiod. J Reprod Fert 65:439–446.

Karsch FJ (1980) Seasonal reproduction: a saga of reversible fertility. The Physiologist 23:29–38.

Karsch FJ, Goodman RL, Legan SJ (1980) Feedback basis of seasonal breeding: test of an hypothesis. J Reprod Fert 58:521–535.

Karsch FJ, Bittman EL, Foster DL, Goodman RL, Legan SJ, Robinson JE (1984) Neuroendocrine basis of seasonal reproduction. Rec Prog Horm Res 40:185–232.

Kennaway DJ, Frith RG, Phillipou G, Matthews CD, Seamark RF (1977) A specific radioimmunoassay for melatonin in biological tissue and fluids and its validation by gas chromatography mass spectrometry. Endocrinology 101:119–127.

Kennaway DJ, Obst JM, Dunstan EA, Friesen HG (1981) Ultradian and seasonal rhythms in plasma gonadotropins, prolactin, cortisol, and testosterone in pinealectomized rams. Endocrinology 108:639–646.

Kennaway DJ, Dunstan EA, Gilmore TA, Seamark RF (1982a) Effects of shortened daylength and melatonin treatment on plasma prolactin and melatonin levels in pinealectomized and sham-operated ewes. Anim Reprod Sci 5:287–294.

Kennaway DJ, Gilmore TA, Seamark RF (1982b) Effects of melatonin implants on the circadian rhythm of plasma melatonin and prolactin in sheep. Endocrinology 110:2186–2188.

Kennaway DJ, Gilmore TA, Seamark RF (1982c) Effect of melatonin feeding on serum prolactin and gonadotropin levels and the onset of seasonal estrous cyclicity in sheep. Endocrinology 110:1766–1772.

Kennaway DJ, Sanford LM, Godfrey B, Friesen HG (1983) Patterns of progesterone, melatonin and prolactin secretion in ewes maintained in four different photoperiods. J Endocrinol 97:229–242.

Kennaway DJ, Dunstan EA, Gilmore TA, Seamark RF (1984) Effects of pinealectomy, oestradiol and melatonin on plasma prolactin and LH secretion in ovariectomized sheep. J Endocrinol 102:199–207.

Legan SJ, Karsch FJ (1980) Photoperiodic control of seasonal breeding in ewes: modulation of the negative feedback action of estradiol. Biol Reprod 23:1061–1068.

Legan SJ, Karsch FJ (1983) Importance of retinal photoreceptors to the photoperiodic control of seasonal breeding in the ewe. Biol Reprod 29:316–325.

Legan SJ, Winans SS (1981) The photoneuroendocrine control of seasonal breeding in the ewe. Gen Comp Endocrinol 45:317–328.

Legan SJ, Foster DL, Karsch FJ (1977) The endocrine control of seasonal reproductive function in the ewe: a marked change in response to the negative feedback action of estradiol on luteinizing hormone secretion. Endocrinology 101:818–824.

Levine JE, Pau K-YF, Ramirez VD, Jackson GD (1982) Simultaneous measurement of luteinizing hormone-releasing hormone and luteinizing hormone release in unanesthetized, ovariectomized sheep. Endocrinology 111:1449–1455.

Lincoln DW, Fraser HM, Lincoln GA, Martin GB, McNeilly AS (1985) Hypothalamic pulse generators. Rec Prog Horm Res 41:369–411.

Lincoln GA (1978) Induction of testicular growth and sexual activity in rams by a "skeleton" short-day photoperiod. J Reprod Fert 52:179–181.

Lincoln GA (1979) Photoperiod control of seasonal breeding in the ram: participation of the cranial sympathetic nervous system. J Endocrinol 82:135–147.

Lincoln GA, Almeida OFX (1982) Inhibition of reproduction in rams by long daylengths and the acute effect of superior cervical ganglionectomy. J Reprod Fert 66:417–423.

Lincoln GA, Davidson DW (1977) The relationship between sexual and aggressive behavior and pituitary and testicular activity during the seasonal sexual cycle of rams, and the influence of photoperiod. J Reprod Fert 49:267–276.

Lincoln GA, Ebling FJP (1985) Effect of constant-release implants of melatonin on seasonal cycles in reproduction, prolactin secretion and moulting in rams. J Reprod Fert 73:241–253.

Lincoln GA, Short RV (1980) Seasonal breeding: Nature's contraceptive. Rec Prog Horm Res 36:1–52.

Lincoln GA, Peet MJ, Cunningham RA (1977) Seasonal and circadian changes in the episodic release of follicle-stimulating hormone, luteinizing hormone and testosterone in rams exposed to artificial photoperiods. J Endocrinol 72:337–349.

Lincoln GA, Almeida OFX, Arendt J (1981) The role of melatonin and circadian rhythms in seasonal reproduction in rams. J Reprod Fert Suppl 30:23–31.

Lincoln GA, Almeida OFX, Klandorf H, Cunningham RA (1982) Hourly fluctuations in the blood levels of melatonin, prolactin, luteinizing hormone, follicle-stimulating hormone, testosterone, triiodothyronine, thyroxine and cortisol in rams under artificial photoperiods, and the effects of cranial sympathectomy. J Endocrinol 92:237–250.

Marshall FHA (1937) On changeover in oestrous cycle in animals after transference across equator, with further observations on incidence of breeding season and factors controlling sexual periodicity. Proc R Soc Lond Ser B 122:413–428.

Martin GB, Scaramuzzi RJ (1983) The induction of oestrus and ovulation in seasonally anovular ewes by exposure to rams. J Steroid Biochem 19:869–875.

Martin GB, Scaramuzzi RJ, Henstridge JD (1983) Effects of oestradiol, progesterone and androstenedione on the pulsatile secretion of luteinizing hormone in ovariectomized ewes during spring and autumn. J Endocrinol 96:181–193.

Mauleon P, Rougeot J (1962) Régulation des saisons sexuelles chez des brebis de races différentes au moyen de divers rhythmes lumineux. Ann Biol Anim Biochim Biophys 2:209–222.

Meyer SL, Goodman RL (1985) Neurotransmitters involved in mediating the steroid-dependent suppression of pulsatile luteinizing hormone secretion in anestrous ewes: effects of receptor antagonists. Endocrinology 116:2054–2061.

Meyer SL, Goodman RL (1986) Separate neural systems mediate the steroid-dependent and steroid-independent suppression of tonic luteinizing hormone secretion in the anestrous ewe. Biol Reprod 35:562–571.

Montgomery GW, Martin GB, Pelletier J (1985) Changes in pulsatile LH secretion after ovariectomy in Ile-de-France ewes in two seasons. J Reprod Fert 73:173–183.

Nett TM, Niswender GD (1982) Influence of exogenous melatonin on seasonality of reproduction in sheep. Theriogenology 17:645–653.

Newton JE, Betts JE (1972) A comparison between the effects of various photoperiods on the reproductive performance of Scottish Halfbred ewes. J Agric Sci 78:425–433.

Pau K-YF, Jackson GL (1985) Effect of frontal hypothalamic deafferentation on photoperiod-induced changes in luteinizing hormone secretion in the ewe. Neuroendocrinology 41:72–78.

Pau K-YF, Kuehl DE, Jackson GL (1982) Effect of frontal hypothalamic deafferentation on luteinizing hormone secretion and seasonal breeding in the ewe. Biol Reprod 27:999–1009.

Pelletier J, Ortavant R (1975a) Photoperiodic control of LH release in the ram I. Influence of increasing and decreasing light photoperiods. Acta Endocrinol (Kbh) 78:435–441.

Pelletier J, Ortavant R (1975b) Photoperiodic control of LH release in the ram II. Light-androgen interaction. Acta Endocrinol (Kbh) 78:442–450.

Przekop F (1978) Effect of anterior deafferentation of the hypothalamus on the release of luteinizing hormone (LH) and reproductive function in sheep. Acta Physiol Pol 29:393–407.

Przekop F, Domanski E (1980) Abnormalities in the seasonal course of oestrous cycles in ewes after lesions of the suprachiasmatic area of the hypothalamus. J Endocrinol 85:481–486.

Radford HM (1961) Photoperiodism and sexual activity in Merino ewes. I. The effect of continuous light on the development of sexual activity. Aust J Agric Res 12:139–146.

Reiter RJ (1972) Evidence for photorefractoriness of the pituitary gonadal axis to the pineal gland and its possible implication in annual reproductive cycles. Anat Rec 173:365–371.

Reiter RJ (1980) The pineal and its hormones in the control of reproduction in mammals. Endocr Rev 1:109–131.

Robinson JE, Karsch FJ (1984) Refractoriness to inductive day length terminates the breeding season of the Suffolk ewe. Biol Reprod 31:656–663.

Robinson JE, Radford HM, Karsch FJ (1985a) Seasonal changes in pulsatile luteinizing hormone (LH) secretion in the ewe: relationship of frequency of LH pulses to day-length and response to estradiol negative feedback. Biol Reprod 33:324–334.

Robinson JE, Wayne NE, Karsch FJ (1985b) Refractoriness to inhibitory day length initiates the breeding season of the Suffolk ewe. Biol Reprod 32:1024–1030.

Robinson TJ (1959) The estrous cycle of the ewe and doe. In: Cole HH, Cupps PT (eds) Reproduction in Domestic Animals. Academic Press, New York, pp. 291–333.

Roche JF, Karsch FJ, Foster DL, Takayi S, Dziuk DJ (1970) Effect of pinealectomy on estrus, ovulation and luteinizing hormone in ewes. Biol Reprod 2:251–254.

Rollag MD, Niswender GD (1976) Radioimmunoassay of serum concentrations of melatonin in sheep exposed to different lighting regimens. Endocrinology 98:482–489.

Rollag MD, O'Callaghan PL, Niswender GD (1978) Serum melatonin concentrations during different stages of the annual reproductive cycle in ewes. Biol Reprod 18:279–285.

Schillo KK, Duehl D, Jackson GL (1985) Do endogenous opioid peptides mediate the effects of photoperiod on release of luteinizing hormone and prolactin in ovariectomized ewes? Biol Reprod 32:779–787.

Stetson MH, Tate-Ostroff B (1981) Hormonal regulation of the annual reproductive cycle of golden hamsters. Gen Comp Endocrinol 45:329–344.

Stetson MH, Watson-Whitmyre M (1976) Nucleus suprachiasmaticus: the biological clock in hamsters? Science 191:197–199.

Tamanini C, Crowder ME, Nett TM (1986) Effects of oestradiol and progesterone on pulsatile secretion of luteinizing hormone in ovariectomized ewes. Acta Endocrinol (Kbh) 111:172–178.

Thimonier J, Brieu V, Ortavant R, Pelletier J (1985) Daylength measurement in sheep. Biol Reprod 32 (Suppl. 1):55.

Thwaites CJ (1965) Photoperiodic control of breeding activity in the Southdown ewe with particular reference to the effects of an equatorial light regime. J Agric Sci 65:57–64.

Turek FW, Campbell CS (1979) Photoperiodic regulation of neuroendocrine-gonadal activity. Biol Reprod 20:32–50.

Turek FW, Swann J, Earnest DJ (1984) Role of the circadian system in reproductive phenomena. Rec Prog Horm Res 40:143–177.

Wodzicka-Tomaszewska M, Hutchinson JDA, Bennett JW (1967) Control of the annual rhythm of breeding in ewes: effect of an equatorial daylength with reversed thermal seasons. J Agric Sci 68:61–67.

Wright PJ, Geytenbeek PE, Clarke IJ, Findlay JK (1981) Evidence for a change in oestradiol negative feedback and LH pulse frequency in post-partum ewes. J Reprod Fert 61:97–102.

Yeates NTM (1949) The breeding season of the sheep with particular reference to its modification by artificial means using light. J Agric Sci 39:1–43.

Yellon SM, Bittman EL, Lehman MN, Olster DH, Robinson JE, Karsch FJ (1985) Importance of duration of nocturnal melatonin secretion in determining the reproductive response to inductive photoperiod in the ewe. Biol Reprod 32:523–529.

Chapter 10

Photoperiodism and Seasonality in Hamsters: Role of the Pineal Gland

Bruce D. Goldman and Jeffrey A. Elliott

I. Pineal and Photoperiodism

During the past two decades, a large body of evidence has been accumulated to indicate a central role for the pineal gland in the regulation of photoperiodic responses in mammals (Goldman and Darrow 1983). Most of this evidence has come from studies of the effects of pinealectomy on the reproductive system in photoperiodic mammals. The Syrian hamster has been the most intensively studied of these species, and frequently has been presented as a "model" for the role of the pineal in regulating reproductive activity. This species is a long-day breeder and exhibits gonadal regression following several weeks of exposure to short days—i.e., daylengths of less than 12.5-h illumination/24-h cycle (Elliott 1976). Removal of the pineal completely prevents this response to short-day exposure; pinealectomized Syrian hamsters remain reproductively active under all photoperiodic conditions (Reiter 1969, Reiter 1974). The photoperiod dependency of the response to pinealectomy in this species has been emphasized. In long days, pinealectomized and intact Syrian hamsters show identical levels of reproductive activity; the effect of pinealectomy becomes apparent only during exposure to short days (Reiter 1974). The closely related Turkish hamster is also a long-day breeder and requires daylengths of nearly 16 h to stimulate reproductive activity (Hong et al. 1986). Pinealectomy also has a photoperiod-dependent effect on reproductive activity in this species, but the nature of the response to pinealectomy is almost the reverse of that seen in Syrian hamsters. In Turkish hamsters, pinealectomy actually triggers testicular regression in long-day-housed animals; short-day-exposed Turkish hamsters exhibit gonadal regression regardless of whether or not they have been pinealectomized (Carter et al. 1982).

Responses to pinealectomy in other photoperiodic species are often somewhat "intermediate" between the extremes represented by Syrian and Turkish hamsters, respectively. In Djungarian hamsters, the reproductive state which follows pinealectomy depends upon prior photoperiodic experience. Males which are pinealectomized during exposure to long days remain reproductively active indefinitely, even following transfer to short days, and in this respect are similar to Syrian hamsters. However, when Djungarian hamsters are pinealectomized after testicular regression has occurred in short days, the testes remain regressed for several weeks. Djungarian hamsters, like other long-day breeders, can be photostimulated by exposure to long days following gonadal regression in short photoperiod. If pinealectomy is performed just before transfer from short days to long days, the photostimulatory effect of increased daylength is prevented—i.e., in this paradigm the pineal appears to be a part of the mechanism by which long days stimulate the reproductive axis, as in the Turkish hamster (Hoffmann 1978, Carter and Goldman 1983b).

While these species differences in response to pinealectomy have led to some confusion and to various interpretations of pineal function, there is one feature common to all species examined to date. That is, removal of the pineal gland always appears to interfere with the animal's ability to discriminate between long and short daylengths experienced subsequent to the operation. Syrian hamsters remain reproductively active after pinealectomy regardless of photoperiod; Turkish hamsters exhibit a cycle of gonadal regression followed by recrudescence after pinealectomy in either long or short days. The gonadal state of Djungarian hamsters after pinealectomy depends upon the photoperiod experienced prior to removal of the gland but is not influenced by the photoperiod which is imposed after pinealectomy. These observations suggest that the pineal gland is involved in the transmission of a photoperiodic message to the reproductive system; without the pineal, the reproductive apparatus is unable to respond appropriately to changes in daylength (Goldman 1983).

II. Pineal Melatonin Rhythm: Importance of Peak Duration

A large body of evidence indicates that melatonin is a pineal hormone which is responsible for the control which the gland imposes on the reproductive system in photoperiodic species (Goldman 1983). Perhaps the most notable features of melatonin secretion in mammals are: (1) the dominant circadian rhythmicity which is expressed and (2) the fact that in virtually all mammals studied, pineal melatonin synthesis and secretion occurs predominantly at night, regardless of whether the species is diurnal or nocturnal with respect to general locomotor activity (Goldman and Darrow 1983). Characteristics of the circadian rhythm of pineal melatonin secretion which might vary with the photoperiod include the phase, duration, and amplitude of the nocturnal melatonin peak, as well as the total amount of melatonin secreted. The importance of these various parameters was tested in a series of experiments in juvenile male Djungarian hamsters. Animals were pinealectomized and then administered daily, subcutaneous infusions of melatonin for the next 12 days. The duration of the daily infusions was varied between experimental groups, as was the time of day at which the infusions were administered. In some cases, the amount of melatonin given was also varied between groups by using different concentrations of melatonin in the infusate. This would be expected to result in variations in the amplitude of the serum melatonin peak produced during infusion. The data collected in these experiments suggest that it is the duration of the daily melatonin peak which carries photoperiodic information. The phase of the circadian cycle at which melatonin is secreted, while clearly under strong circadian control (Darrow and Goldman 1986), does not appear to be critical for the photoperiodic response of the reproductive system (Carter and Goldman 1983a,b, Goldman et al. 1984). Also, the results of the infusion studies indicated that in order to elicit a "short-day" response, the duration of the daily peak of melatonin must not only exceed 6 h, but the peak must be relatively continuous for that period of time. Thus, two 5-h infusions, separated from each other by an interval of 2 h, failed to inhibit testicular growth (Figure 10-1).

III. Circadian Regulation of Melatonin Peak Duration

Several types of experiments have revealed the important role of the circadian system in photoperiodic time measurement (i.e., measurement of daylength) in mammals (Elliott and Goldman 1981, Elliott 1981). Thus, it is appropriate to examine the relationship between the circadian system and the pineal gland with respect to the regulation of photoperiodic responses. Previous studies in Djungarian and Syrian hamsters and in laboratory rats had demonstrated the important role of one or more circadian oscillators in regulating the circadian *phase* of melatonin secretion (Illnerova and Vanacek 1982, Tamarkin et al. 1980, Yellon et al. 1982, Elliott and Tamarkin 1987). In view of the results of the infusion studies described above, it became of interest to determine whether the *duration* of the daily melatonin peak is also regulated by a circadian oscillator(s). Juvenile male Djungarian hamsters were exposed to T-cycles with T = 24.33 h and T = 24.78 h, respectively (i.e., 1 h light followed by either 23.33 h or 23.78 h darkness). Examination of locomotor activity records indicated that in both T-cycles, most of the hamsters exhibited a stable pattern of entrainment. Testicular growth was inhibited in T24.33 but was stimulated by T24.78. Serum and pineal melatonin increased precipitously beginning at 2 to 3 h after the termination of the 1-h light pulse in both T-cycles. In T24.78 (photostimulatory), serum and pineal melatonin concentrations remained elevated for about six hours, while in T24.33 (photoinhibitory), melatonin concentrations were elevated for approximately ten hours (Figure 10-2). These results suggest that the *duration* of the nocturnal melatonin peak is regulated largely by one or more circadian oscillators. In addition, the results provide further correlative evidence of the association between long-duration melatonin peaks and inhibition of reproductive development in this species (Darrow and Goldman 1986).

In addition to a lengthened duration of the pineal melatonin peak, entrainment to the T24.78 cycle, as compared to the T24.33 cycle, was associated with an increase in activity time (α), the fraction of the circadian activity cycle characterized by vigorous wheel-running behavior. This compares with the observation that α is longer in entrainment to a shorter photoperiod (e.g., 10L:14D) than it is on a long photoperiod such as 16L:8D. Thus, the effect of T-cycle period on α is a further indication that the circadian system responds to the

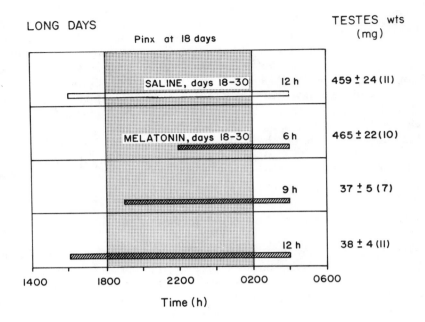

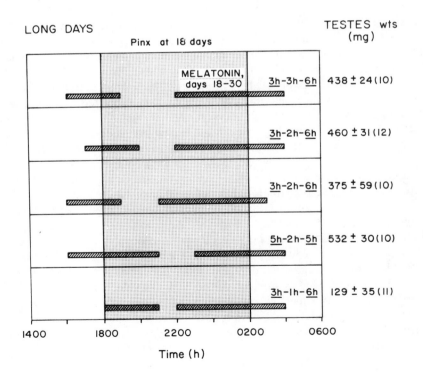

T24.33 cycle as a short day, while it responds to the T24.78 cycle as a long day.

Data suggesting a correlation between long-duration melatonin peaks and inhibition of gonadal function have also been obtained in the Syrian hamster, but direct evidence comparable to the melatonin infusion studies carried out in the Djungarian hamster is not available. Thus in Syrian hamsters, lengthening the photoperiod to 16L:8D led to a reduction in melatonin peak duration as compared to the duration observed on 14L:10D, while the melatonin peak was virtually eliminated in hamsters maintained on 20L:4D. Conversely, several studies have now shown that exposure to shorter daily photoperiods (<10 h), which promote gonadal regression, results in melatonin peak durations that are measurably longer than the duration observed under a 14-h photoperiod (Tamarkin et al. 1980, Rollag et al. 1980, Roberts et al. 1985).

Although it might be supposed that the shorter duration of the nocturnal melatonin peak under long photoperiods is the result of direct inhibition by light, this appears not to be the case. Instead, in the Syrian hamster, like the Djungarian hamster, recent findings indicate that the duration of the pineal melatonin peak is regulated by one or more circadian oscillators. The evidence for this comes from observations of the pineal melatonin profile in Syrian hamsters whose circadian clocks were allowed to freerun in constant darkness (DD) or in continuous, low-intensity (<0.5 lux) red light for up to several months (Tamarkin et al. 1980, Elliott and Tamarkin 1987). Under these conditions, a robust circadian rhythm of pineal melatonin content was observed, and it was found that the pineal rhythm remained in phase with the circadian rhythm of wheel-running activity even after 27 weeks of DD exposure (Figure 10-3). In-

◁————————————————————————————————

Figure 10-1. (A) Effects of timed daily melatonin infusions on testicular development in juvenile hamsters. Males were pinealectomized (Pinx) at 18 days of age and received programmed melatonin infusion from day 18 until day 30. The stippled area represents the dark period of the daily cycle. Times of daily infusion of saline (open bar) or melatonin (hatched bars) are shown along with the duration (6, 9, or 12 h) of infusion for each treatment group. Final mean testes weights with SEM and number of subjects (in parentheses) are shown at the right. (B) Effects of timed daily skeleton melatonin infusions on testicular development in juvenile hamsters. Males were pinealectomized at 18 days of age and received skeleton melatonin infusions from day 18 until day 30. The stippled area represents the dark period of the daily cycle. Times of melatonin infusion are shown (hatched bars). The infusion paradigm is indicated after the second hatched bar. For example, 3h-3h-6h indicates that on each day the animals received a 3-h melatonin infusion, followed by 3 h without treatment, followed by a 6-h melatonin infusion. Groups represented in the top two panels and the bottom panel received 5 ng melatonin during each 3-h infusion and 10 ng melatonin during each 6-h infusion. The group represented in the middle panel received 10 ng melatonin during the 3-h infusion and 20 ng melatonin during the 6-h infusion. The remaining group was administered 10 ng melatonin during each 5-h infusion. Final mean testes weights with SEM and number of subjects (in parentheses) are shown at the right. (From Goldman et al., © 1984 The Endocrine Society. Reprinted with permission.)

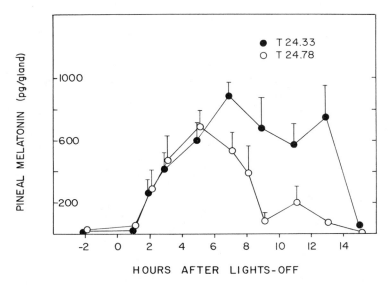

Figure 10-2. Pineal melatonin content in male hamsters exposed for 1 month to non-photostimulatory (T24.33) or photostimulatory (T24.78) T-cycles. Pineals were obtained at different times of day relative to the end of the 1-h light pulse (hour zero). Each point represents the mean of 6 to 8 individual glands. Where standard errors are not shown, the value did not exceed the radius of the symbol. (From Darrow and Goldman, 1986.)

terestingly, although the "evening" rise of pineal melatonin content occurred at approximately the same circadian time, relative to the onset of wheel-running, regardless of the length of DD exposure, the "morning" decline in pineal melatonin content was delayed 4 to 5 h by prolonged DD exposure (Elliott and Tamarkin 1987). In a second set of experiments, the clock resetting effects of light (15-min pulses) were studied in order to obtain additional evidence pertaining to circadian regulation of pineal melatonin. It was found that light pulses which elicited phase advances in the circadian wheel-running rhythm also advanced the phase of the pineal melatonin peak in subsequent cycles of a DD freerun (Figure 10-4). These experiments also showed a strong correlation between compression of α following the phase-advancing stimulus and a parallel decrease in the duration of the pineal melatonin peak, both of which persisted for several cycles in the absence of any further exposure to light. The transient decrease in melatonin peak duration persisting for several days following a single phase-shifting stimulus, associated with different rates of phase-shifting for the "evening" rise and "morning" decline in pineal melatonin content, suggests that the timing of these two events is governed by two or more circadian oscillators. Thus, it appears that in the Syrian hamster both melatonin peak duration and α may be regulated by the phase angle difference among two or more circadian oscillators which comprise a complex circadian pacemaker (Pittendrigh and Daan 1976b). In this context, it is noteworthy that Elliott (1981) found that

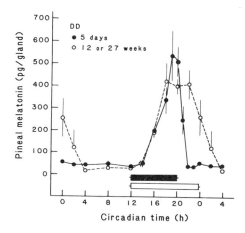

Figure 10-3. Circadian profiles of pineal melatonin content (PMEL) and wheelrunning activity of Syrian hamsters maintained for varied lengths of time in continuous darkness (DD). Individual data were averaged over 1–2-h bins of circadian time and each point plots the mean ± SE for a group of 4 to 17 hamsters. Closed circles: pooled data from two experiments. In each experiment, activity was recorded from a subgroup of 4 to 5 animals and the circadian time of all individual pineal samples was calculated using the mean projected time of activity onset of the group. Open circles: activity was recorded from all animals and the circadian time of each sample was calculated individually. Data from animals studied at 12 and 27 weeks are pooled. Bars represent the duration of activity time (α, mean ± SE) after 5 days (closed bar) and after 12 or 27 weeks (open bar) in DD. Prolonged DD resulted in increases of 4 to 5 h in both the duration of the melatonin peak (= hours PMEL > 125 pg/gland) and in the duration of α. (From Elliott and Tamarkin, 1987.)

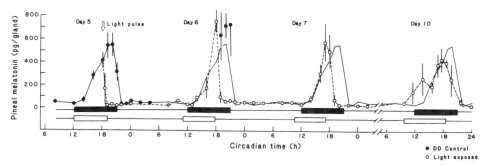

Figure 10-4. Phase advance of the circadian rhythm of pineal melatonin content in the Syrian hamster. A 15-min light pulse occurring 6 h after activity onset (CT18) causes an immediate suppression of pineal melatonin. In subsequent cycles, peak duration is transiently reduced due to a 3-h phase advance in the fall of PMEL to baseline accompanied by little change in the timing of the rise. At 5 days after the light pulse (Day 10), both the rising and falling phase of the PMEL rhythm were advanced, restoring melatonin peak duration to approximately its initial (Day 5) value (see text and Figure 10-3 for details). (From Elliott and Tamarkin, 1987.)

light pulses which elicited large phase shifts in wheel-running activity also delayed gonadal regression in DD. Interestingly, light pulses which elicited large phase advances followed for several days by a pronounced compression of α were most effective in blocking gonadal regression. Thus, in the Syrian hamster there is ample correlative evidence that photoperiodic treatments which lead to gonadal regression are associated with long-duration melatonin peaks and that light treatments which maintain gonadal function are associated with shorter-duration melatonin peaks.

IV. Evolution of Photoperiodism in Mammals

Pineal involvement in photoperiodic responsiveness has now been examined in a wide variety of photoperiodic mammals representing seven different orders, including the Marsupialia. In virtually all the photoperiodic mammals which have been examined, pinealectomy or other procedures known to interfere with pineal function seriously interfered with the ability to respond to photoperiod changes (Goldman 1983). In contrast, pinealectomy or melatonin administration was relatively ineffective in altering reproductive state in "nonphotoperiodic" species, such as adult laboratory rats and house mice (Turek et al. 1976, Goldman et al. 1981). These observations suggest that a central and specific role of the pineal gland in mammalian photoperiodism was probably established very early in mammalian evolution, if not earlier. Important questions remain, however, for the evolution of pineal function in mammals: How many mammals are photoperiodic? If a significant percentage of mammalian species are nonphotoperiodic, does the pineal have some other function in these species?

It may be that the number of photoperiodic species has been underestimated as a result of overemphasis on the role of photoperiod in reproductive strategy as compared to other types of seasonal changes in physiology and behavior. Animals, such as laboratory rats and mice, which show little or no effect of daylength on reproductive activity, have frequently been considered to be nonphotoperiodic. However, in other species, several additional parameters have been found to be largely under photoperiodic control. Examples of such characteristics are seasonal pelage changes in snowshoe hares, weasels, mink, white-footed mice, and Djungarian hamsters (Lyman 1943, Lynch and Epstein 1976, Rust and Meyer 1969, Martinet et al. 1981), body weight and body fat stores in Syrian hamsters and Djungarian hamsters (Bartness and Wade 1985), and thermoregulatory adaptations in Djungarian hamsters and white-footed mice (Lynch and Epstein 1976, Steinlechner and Heldmaier 1982). In several cases, these "nonreproductive" photoperiodic responses have been shown to occur partially or totally independently of changes in gonadal hormones; for example, castrated Djungarian hamsters undergo photoperiod-induced changes in body weight, pelage characteristics, and thermoregulation which are virtually identical to those observed in gonad-intact hamsters (Vitale et al. 1985). Recent studies in one supposedly "nonphotoperiodic" species, the laboratory rat, suggest that daylength does influence the amount of brown fat growth during early life (Vanacek and Illnerova 1985).

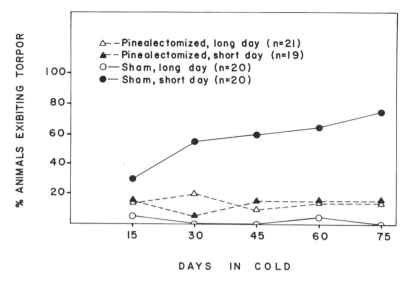

Figure 10-5. Effects of photoperiod and pinealectomy on percent of castrated hamsters exhibiting daily torpor. The groups exposed to short days had been exposed to a short-day warm environment for 8 weeks before transfer to the short-day cold room. Each point represents the percent of animals in the group which displayed at least 1 day of torpor during the preceding 15-day period. (From Vitale et al. 1985.)

In most cases that have been examined, "nonreproductive" responses to photoperiod are prevented by pinealectomy in much the same manner as are the photoperiodic reproductive changes. For example, in the Djungarian hamster, daily torpor was observed only in animals exposed to short days; long-day hamsters failed to exhibit daily torpor even after castration. Pinealectomized hamsters showed a very low incidence of torpor in either long or short days (Figure 10-5). As with this thermoregulatory response, the effects of daylength on body weight and pelage also appear to require the participation of the pineal gland (Vitale et al. 1985), and melatonin has been implicated in all these responses (Steinlechner and Heldmaier 1982, Hoffmann 1978). It seems likely that the circadian pattern of pineal melatonin secretion acts as a mediator of the photoperiodic signal for these responses similarly to its action with respect to reproductive responses. If this is true, then studies of the evolution of photoperiodism in mammals should include a variety of responses under the definition of "photoperiodism," based on the probability that the different response systems share a common neuroendocrine mechanism for the processing and transmission of daylength information. The case for a central role of the pineal gland in mammalian photoperiodism can be summarized as follows:

1. Most of the physiological functions of the mammalian pineal that have been clearly defined involve photoperiodic responses.
2. Most mammalian photoperiodic responses that have been carefully examined appear to involve the pineal gland.

V. Effects of Photoperiod on the Circadian Pacemaker

The first studies to indicate that circadian rhythms are involved in mammalian photoperiodism were based on concurrent analysis of the reproductive responses and the entrainment of the circadian wheel-running rhythms of male Syrian hamsters subjected to different photoperiods and to a variety of non-24-h T-cycles (Elliott 1976). These experiments showed that the reproductive response to light is closely correlated with the pattern of circadian entrainment to the LD cycle: entrainment to a short photoperiod (1–6 h) is photosimulatory only in those T-cycles in which light coincides with the active phase of the circadian wheel-running rhythm. Thus, the circadian time of light exposure proved to be far more important than the duration of the photoperiod per se.

The mechanism by which seasonal alteration in daylength is encoded as change in the circadian pattern of pineal melatonin secretion must involve entrainment of the circadian pacemaker which ultimately regulates melatonin secretion to the light–dark cycle. It is possible to describe the entrainment of a circadian oscillator to the LD cycle as a function of the oscillator's phase response curve (PRC) for light stimuli (Pittendrigh 1965, 1981). The PRC describes a circadian rhythm of resetting response to light which is thought to be an essential component of circadian pacemaker physiology and one which directly mediates synchronization of the circadian system to the daily light–dark cycle (Pittendrigh 1965). Graphically, it represents the phase-shifting responses elicited by a standard entraining stimulus (e.g., 15 min fluorescent light at 100 lux) experienced at different phases of a circadian rhythm, studied under freerunning conditions (e.g., continuous darkness). It has been well established that the shape and amplitude characteristics of the PRC vary with the strength and duration of the stimulus and also depending on the species (Aschoff 1965, Takahashi et al. 1984). In addition, Figure 10-6 shows that in the Syrian hamster these characteristics of the PRC are also strongly dependent on the photoperiod to which the animals were previously entrained (Elliott and Pittendrigh 1987). Following entrainment to a short photoperiod (10L:14D), phase-shift responses occur over a broader range of the circadian cycle and the amplitude of the PRC is increased. In view of (1) the established role of melatonin in regulating seasonal physiological changes and (2) the accumulating evidence for the importance of circadian regulation of the *duration* of pineal melatonin secretion, these findings have important implications. Photoperiodic induction of seasonally appropriate changes in mammalian physiology may involve changes in the properties of the circadian pacemaker itself, including alteration of its entrainment response. Additionally, change in the time course of the pacemaker's circadian cycle, as reflected in the waveform of its PRC, may directly underlie the durational changes in activity time and melatonin secretion which occur downstream. Because photoperiodically regulated changes in pacemaker physiology persist for many cycles in DD and are known to influence entrainment for many weeks following a photoperiod change (Pittendrigh and Daan 1976a), it appears likely that such changes in the pacemaker may play a significant role in photoperiodic responses which are mediated by the circadian system and the pineal

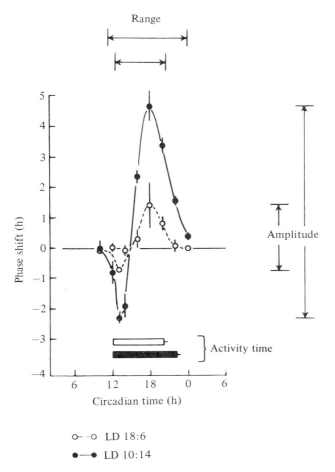

Figure 10-6. Photoperiod dependence of the Syrian hamster's phase response curve (PRC). Each point plots the mean phase shift ($\triangle\phi \pm$ SE) for a group of 3 to 5 hamsters pulsed with light (15 min) at the indicated circadian time (CT) during the tenth day of constant darkness (DD) following release from entrainment to the LD cycle. Open symbols: long photoperiod (18L:6D); closed symbols: short photoperiod (10L:14D). Bars represent the duration of nocturnal wheel-running (mean $\alpha \pm$ SE) with onset of activity designated at CT12. Following the short photoperiod (10L:14D), phase-shift responses (($\triangle\phi \geqslant 0.5$ h) occur over a broader range of the circadian cycle (the "dead zone" is reduced) and the PRC has a larger amplitude than it does following entrainment to the long photoperiod (18L:6D). The decreased range and amplitude of PRC after 18L:6D is correlated with a decrease in activity time (α), and both observations reflect the photoperiod dependence of the physiological state of a complex circadian pacemaker. (From Pittendrigh et al., © 1984, John Wiley and Sons Ltd. Reprinted with permission.)

gland. Finally, it is noteworthy that the ability of the circadian pacemaker to be influenced by previous photoperiodic history has the potential of providing an added degree of flexibility in responses to the annual cycle of continually changing daylengths (see below).

VI. Use of Photoperiod as Continuous Reference Cue for Seasonality

The terms "long days" and "short days" have been used very extensively in the description of photoperiodic phenomena. The use of these terms may sometimes be taken to imply that for a given species, or at least for a given type of photoperiodic response in a particular species, daylengths fall neatly into one of the two categories. In the Syrian hamster, for example, daylengths of less than 12.5 hours are considered short days because reproduction is inhibited in such daylengths; conversely, daylengths of greater than 12.5 hours are classified as long days. Recent observations suggest that this rigid classification of daylengths as "long" or "short" may break down in certain situations. Two examples serve to illustrate the nature of the challenges to this simple classification:

1. When adult Djungarian hamsters which have been raised from birth in 16L:8D are transferred to 14L:10D, testicular regression ensues. In standard terms of reference then, 14L:10D would be classified as a "short day" for the Djungarian hamster. However, exposure to 14L:10D does not induce a molt to the winter pelage, another "short-day" response in this species; the development of the winter coat requires yet shorter photoperiods (Duncan et al. 1985).
2. In addition, 14L:10D cannot be unequivocally designated as a "short day" even for the reproductive response in the Djungarian hamster. Thus, when adult male hamsters were first exposed to 8L:16D to induce testicular regression, subsequent transfer to 14L:10D actually stimulated gonadal recrudescence (Hoffmann 1984). Therefore, it appears that the nature of the reproductive response to 14L:10D depends on prior photoperiodic history.

A similar dependence of reproductive response on prior photoperiodic history has been demonstrated in experiments designed to examine the ability of rodents to receive photoperiodic information during fetal life. When pregnant Montane voles (Horton 1985) or Djungarian hamsters (Stetson et al. 1986) were exposed to 16L:8D throughout gestation and were transferred, along with their offspring, to 14L:10D beginning on the day of parturition, the young males exhibited relatively slow testicular growth. However, when gestation occurred in 8L:16D (voles) or 12L:12D (hamsters) and the litters were transferred to 14L:10D at birth, testicular growth occurred more rapidly. Cross-fostering experiments in both species have demonstrated that photoperiodic information is received by the developing males before birth, probably having been transmitted in some way from mother to fetus (Horton 1985, Reppert et al. 1985). This may represent an adaptive response. A 14L:10D daylength is, by itself, ambiguous in that it might signal either spring or autumn. However, by taking into account whether the daylength is increasing or decreasing, animals may be able to ensure that the appropriate response be made to a 14L:10D photoperiod. By such a mechanism, animals may be able to use photoperiod as a more or less continuous reference cue for determining time of year, despite the fact that most daylengths occur twice during each annual cycle.

VII. Evidence for Endogenous, Long-Interval Timing Mechanisms in Mammals

Endogenous circadian timing mechanisms, or circadian oscillators, appear to be an almost universal attribute of cellular organisms (Pittendrigh 1981). Many organisms also possess endogenous mechanisms which allow them to measure time on a much longer scale. Some mammals exhibit circannual rhythms—i.e., rhythms with a period length close to one year—when maintained under constant conditions in the laboratory (Pengelley 1974). Because of the difficulty of studying such long-term rhythms, we know much less about the phyletic distribution of circannual clocks as compared to what is known about circadian rhythmicity.

Most of the photoperiodic mammals discussed in this paper do not appear to exhibit circannual rhythms when held under constant conditions. However, these species do display a characteristic which suggests that they possess some type of endogenous mechanism which measures time on a seasonal scale. Thus, in every species examined where winter responses are induced by exposure to decreased daylengths, a "spontaneous" return to the summer type condition occurs after several months, even when photoperiodic conditions (i.e., short daylength, or even continuous darkness) remain unchanged. The most widely studied example of this class of phenomena is the so-called "spontaneous recrudescence" of the gonads which has been observed in every species of long-day breeder following several months of exposure to short photoperiod. It has been suggested that this "spontaneous" reversion to a reproductively active state may represent an adaptation which allows the animal to anticipate favorable environmental conditions even before the advent of "long" daylengths (Reiter 1974). If so, one would expect selection pressure to be exerted on the mechanism which triggers gonadal recrudescence so that reproductive capacity would be attained at the most opportune time. Certainly, the synchrony of the recrudescence response among individuals of a species is impressive. It has been suggested, therefore, that the timing of spontaneous recrudescence is regulated by an endogenous timing mechanism (Elliott and Goldman 1981). The annual reproductive cycle in these species might then be considered to be a product of two components: (1) a photoperiodic mechanism which determines (on the basis of the critical photoperiod) when gonadal regression will occur and (2) an endogenous, long-term timing mechanism which is set into motion by exposure to short days and determines the time of gonadal recrudescence. A similar hypothesis could be advanced for short-day breeding species, such as the sheep, where "spontaneous" gonadal recrudescence occurs after a prolonged period of gonadal inhibition in long days (Robinson and Karsch 1984).

It is tempting to speculate that the hypothetical timer which establishes the time for gonadal recrudescence is related to the timing mechanism which serves as the basis for circannual rhythmicity in some mammals. It is possible that a discrete neural "circannual" oscillator may exist in species which exhibit circannual rhythms, though attempts to locate such an oscillator have not yet been successful (Dark and Zucker 1985). If such a neural substrate is present in circannual species, it is possible that the same locus is involved in seasonal timing in all photoperiodic mammals.

It is interesting to consider, in the context of present knowledge, what proportion of mammalian species employ some combination of photoperiodism and "seasonal timing" to maintain a "calendar" which can be used as a basis for time reference in the regulation of annual cycles. Since both photoperiodism and some form of endogenous "seasonal timer" are widespread among mammals, it seems likely that these mechanisms were available to early mammalian ancestors. It would appear that the use of such mechanisms to provide a calendar for continuous reference might offer very important adaptive advantages. If these mechanisms were, in fact, already established in early mammals, then the absence of such mechanisms in any present-day mammal would have to be viewed as resulting from a loss during more recent evolution. This hypothetical view places a somewhat different perspective on mammalian photoperiodism. That is, one may need to ask how likely it would be that continuous access to accurate information on time of year would be of no selective value, resulting in the loss of those physiological mechanisms which provide such information. There are probably few environmental niches where time-of-year information would not provide predictive information of potential adaptive value.

References

Aschoff J (1965) Response curves in circadian periodicity. In: Aschoff J (ed) Circadian Clocks. North-Holland, Amsterdam, pp. 95–111.

Bartness TJ and Wade GN (1985) Photoperiodic control of seasonal body weight cycles in hamsters. Neurosci Behav. Rev. 9:599–612.

Carter DS and Goldman BD (1983a) Antigonadal effects of timed melatonin infusion in pinealectomized male Djungarian hamsters *(Phodopus sungorus sungorus):* duration is the critical parameter. Endocrinology 113:1261–1267.

Carter DS and Goldman BD (1983b) Progonadal role of the pineal in the Djungarian hamster *(Phodopus sungorus sungorus):* mediation by melatonin. Endocrinology 113:1268–1273.

Carter DS, Hall VD, Tamarkin L, Goldman BD (1982) Pineal is required for testicular maintenance in the Turkish hamster *(Mesocricetus brandti)*. Endocrinology 111:863–871.

Dark J, Zucker I (1985) Circannual rhythms of ground squirrels: role of the hypothalamic paraventricular nucleus. J Biol Rhythms 1:17–23.

Darrow JM, Goldman BD (1986) Circadian regulation of pineal melatonin and reproduction in the Djungarian hamster. J Biol Rhythms 1:39–54.

Duncan MJ, Goldman BD, Di Pinto MN, Stetson MH (1985) Testicular function and pelage color have different critical daylengths in the Djungarian hamster, *Phodopus sungorus sungorus*. Endocrinology 116:424–430.

Elliott JA (1976) Circadian rhythms and photoperiodic time measurement in mammals. Fed Proc 35:2339–2346.

Elliott JA (1981) Circadian rhythms, entrainment and photoperiodism in the Syrian hamster. In: Follett BK, Follett DE (eds.) Biological Clocks in Seasonal Reproductive Cycles, John Wright and Sons, Bristol, England, pp. 203–217.

Elliott JA, Goldman BD (1981) Seasonal reproduction: photoperiodism and biological clocks. In: Adler NT (ed) Neuroendocrinology of Reproduction: Physiology and Behavior, Plenum Press, New York, pp. 377–423.

Elliott JA, Pittendrigh CS (1987) After-effects of entrainment on the phase response curve of Syrian hamsters. Manuscript in preparation.

Elliott JA, Tamarkin L (1987) Complex structure of the circadian pacemaker regulating pineal melatonin content and wheel-running activity in the Syrian hamster. Manuscript in preparation.

Goldman BD (1983) The physiology of melatonin in mammals. In: Reiter RJ (ed), Pineal Research Reviews, Vol. 1. Alan R. Liss, Inc., New York, pp. 145–182.

Goldman BD, Darrow JM (1983) The pineal gland and mammalian photoperiodism. Neuroendocrinology 37:386–396.

Goldman BD, Darrow JM, Yogev L (1984) Effect of timed melatonin infusions on reproductive development in the Djungarian hamster *(Phodopus sungorus)*. Endocrinology 114:2074–2083.

Goldman B, Hall V, Hollister C, Reppert S, Roychoudhury P, Yellon S, Tamarkin L (1981) Diurnal changes in pineal melatonin content in four rodent species: relationship to photoperiodism. Biol Reprod 24:778–783.

Hoffman K (1978) Photoperiodic mechanism in hamsters: the participation of the pineal gland. In: Follett BK, Follett DE (eds) Environmental Endocrinology. Springer-Verlag, Berlin, pp. 94–102.

Hoffmann K (1984) Photoperiodic reaction in the Djungarian hamster is influenced by previous light history. Soc Study Reprod 30:(Abstract 50), p. 55.

Hong SM, Rollag MD, Stetson MH (1986) Maintenance of testicular function in Turkish hamsters: interaction of photoperiod and the pineal gland. Biol Reprod 34:527–531.

Horton TH (1985) Cross-fostering of voles demonstrates *in utero* effect of photoperiod. Biol Reprod 33:934–939.

Illnerova H, Vanacek J (1982) Two-oscillator structure of the pacemaker controlling the circadian rhythm of N-acetyltransferase in the rat pineal gland. J Comp Physiol A145:539–548.

Lyman CP (1943) Control of coat color in the varying hare, *Lepus americanus* Erxleben. Bull Museum Comp Zool Harvard 93:393–461.

Lynch GR, Epstein AL (1976) Melatonin induced changes in gonads, pelage, and thermogenic characters in the white-footed mouse, *Peromyscus leucopus*. Comp Biochem Physiol 53C:67.

Martinet L, Meunier M, Allain D (1981) Control of delayed implantation and onset of spring moult in the mink *(Mustela vison)* by daylight ratio, prolactin and melatonin. In: Ortovant R, Pelletier J, Ravault JP (eds) Photoperiodism and Reproduction. INRA Publ., Paris, pp. 253–261.

Pengelley ET (1974) Circannual Clocks. Academic Press, New York.

Pittendrigh CS (1965) On the mechanism of the entrainment of a circadian rhythm by light cycles. In: Aschoff J (ed) Circadian Clocks. North-Holland, Amsterdam, pp. 277–297.

Pittendrigh CS (1981) Circadian organization and photoperiodic phenomena. In: Follett BK, Follett DE (eds) Biological Clocks in Seasonal Reproductive Cycles. John Wright and Sons, Bristol, England, pp. 1–35.

Pittendrigh CS, Daan S (1976a) A functional analysis of circadian pacemakers IV. Entrainment: pacemaker as clock. J Comp Physiol 106:291–331.

Pittendrigh CS, Daan S (1976b) A functional analysis of circadian pacemakers V. Pacemaker structure: a clock for all seasons. J Comp Physiol 106:333–355.

Pittendrigh CS, Elliott J, Takamura T (1984) The circadian component in photoperiodic induction. In: Photoperiodic Regulation of Insect and Molluscan Hormones (Ciba Foundation Symposium 104). Pitman, London, pp. 26–47.

Reiter RJ (1969) Pineal function in long term blinded male and female golden hamsters. Gen Comp Endocrinol 12:460–468.

Reiter RJ (1974) Circannual reproductive rhythms in mammals related to photoperiod and pineal function: a review. Chronobiologia 1:365–395.

Reppert SM, Duncan MJ, Goldman BD (1985) Photic influences on the developing mammal. In: Evered D, Clark S (eds) Photoperiodism, Melatonin and the Pineal (Ciba Foundation Symposium 117). Pitman, London, pp. 116–128.

Robinson JE, Karsch FJ (1984) Refractoriness to inductive day lengths terminates the breeding season of the Suffolk ewe. Biol Reprod 31:656–663.

Roberts AC, Martensz ND, Hastings MH, Herbert J (1985) Changes in photoperiod alter the daily rhythm of pineal melatonin content, hypothalamic β-endorphin content and the LH response to naloxone in the male Syrian hamster. Endocrinology 117:141–147.

Rollag MD, Panke ES, Reiter RJ (1980) Pineal melatonin content in male hamsters throughout the seasonal reproductive cycle. Proc Soc Exp Biol Med 165:330–334.

Rust CC, Meyer RK (1969) Hair color, molt, and testes size in short-tailed weasels treated with melatonin. Science 165:921–922.

Steinlechner S, Heldmaier G (1982) Role of photoperiod and melatonin in seasonal acclimatization of the Djungarian hamster, *Phodopus sungorus*. Int J Biometeor 26:329–337.

Stetson MH, Elliott JA, Goldman BD (1986) Maternal transfer of photoperiodic information influences the photoperiodic response of prepubertal Djungarian hamsters. Biol Reprod 34:664–669.

Takahashi JS, DeCoursey PJ, Bauman L, Menaker M (1984) Spectral sensitivity of a novel photoreceptive system mediating entrainment of mammalian circadian rhythms. Nature 308:186–188.

Tamarkin L, Reppert SM, Klein DC, Pratt BL, Goldman BD (1980) Studies on the daily pattern of pineal melatonin in the Syrian hamster. Endocrinology 107:1525–1529.

Turek FW, Desjardins C, Menaker M (1976) Differential effects of melatonin on the testes of photoperiodic and nonphotoperiodic rodents. Biol Reprod 15:94–97.

Vanacek J, Illnerova H (1985) Effect of short and long photoperiods on pineal N-acetyltransferase rhythm and on growth of testes and brown adipose tissue in developing rats. Neuroendocrinology 41:186–191.

Vitale PM, Darrow JM, Duncan MJ, Shustak CA, Goldman BD (1985) Effects of photoperiod, pinealectomy and castration on body weight and daily torpor in Djungarian hamsters *(Phodopus sungorus)*. J Endocrinol 106:367–375.

Yellon SM, Tamarkin L, Pratt BL, Goldman BD (1982) Pineal melatonin in the Djungarian hamster: photoperiodic regulation of a circadian rhythm. Endocrinology 111:488–492.

Chapter 11

Reproductive Refractoriness in Hamsters: Environmental and Endocrine Etiologies

MARCIA WATSON-WHITMYRE and MILTON H. STETSON

I. Introduction

A. Annual Reproductive Cycle of the Golden Hamster

The golden hamster, *Mesocricetus auratus*, is typical of temperate-latitude rodents in its annual reproductive cycle. When housed under natural photoperiods, adult hamsters are reproductively active in the spring and summer months but inactive during the fall and winter (Vendrely et al. 1971a,b, 1972; see Figure 11-1A). Laboratory experiments have demonstrated that the most important *Zeitgeber* for the annual reproductive cycle is the yearly variation in photoperiod: the entire breeding cycle can be reproduced under artificial daylengths when temperature and nutritional variables are held constant (Reiter 1968a). The critical daylength for the golden hamster is 12.5 hours of light per day: light cycles containing more than 12.5 hours of light are perceived as long days and are favorable to gonadal growth and maintenance, while those shorter than 12.5 hours of light result in gonadal regression (Gaston and Menaker 1967). The reproductive changes brought about by varying the daylength are mediated by changes in gonadotropin release (Berndtson and Desjardins 1974).

The phase of gonadal regression, occurring in the late summer and fall under natural daylengths, requires 8 to 12 weeks, and is thus complete by late October (Vendrely et al. 1972). The reproductive system then enters a quiescent phase lasting another 6 to 10 weeks. Commencing in February under natural photoperiods, the gonads of both sexes begin to grow and reestablish gametogenic activity; this growth is termed *spontaneous recrudescence* because it occurs without the influence of stimulatory daylengths, that is, those greater than 12.5 hours of light per day. Indeed, spontaneous recrudescence can even be demonstrated in blinded hamsters (Reiter 1969a,b); blinding is equivalent to exposure to constant darkness since the only known receptor for photoperiodic information is the retina (Reiter 1968b). Following spontaneous recrudescence, the hamster remains in breeding condition throughout the spring and summer

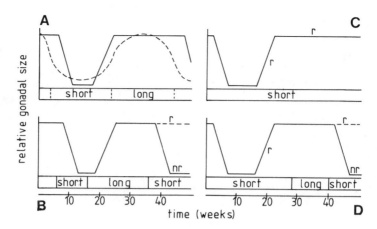

Figure 11-1. Schematic diagrams showing the reproductive response of hamsters to various photoperiodic manipulations, shown by the horizontal bars marked "long" or "short." In (A), the response to natural photoperiods is shown by the solid line; daylength is indicated by the broken line. In (B), the annual reproductive cycle is mimicked by appropriate photoperiodic manipulation in the laboratory. Note that gonadal recrudescence can be stimulated by transfer to long photoperiods when the testes are regressed. If refractoriness (r) has been induced during the initial short-day treatment, the hamsters will not regress when exposed to short days a second time; nonrefractory (nr) hamsters will show a second bout of regression. In (C), spontaneous recrudescence on short days is shown; hamsters may be considered refractory (r) during spontaneous recrudescence, or because of the continued failure to regress a second time. In (D), spontaneous recrudescence is again shown, along with the ability of long days to terminate the refractory condition. If refractory (r) hamsters are exposed to more than 11 weeks of long days, the refractory condition will be terminated and gonadal regression will result when the nonrefractory (nr) hamsters are reexposed to short days. If the long-day exposure is shorter than 11 weeks, the refractory (r) condition will not be terminated.

months. The hamster can also be induced to undergo *photoinduced recrudescence* by treatment with suitable long photoperiods at any time after the commencement of gonadal regression (Figure 11-1B). Both spontaneous and photoinduced recrudescence require about 8 to 10 weeks for completion; thus the complete gonadal cycle, from maximum activity to full quiescence back to maximum activity, takes a total of 20 to 30 weeks in the laboratory. For a complete review of the annual reproductive cycle in the hamster, see Stetson and Tate-Ostroff (1981) or Reiter (1980).

B. Definition of Refractoriness

The hamster undergoing spontaneous recrudescence is, by definition, *refractory* to the prevailing photoperiod, in that the normal response to short photoperiods, namely, gonadal regression, is not seen. However, the period of photorefractoriness is more prolonged than the 10 or so weeks required for the resumption

of full gametogenic activity. Hamsters that have undergone spontaneous recrudescence will not experience a second bout of gonadal regression if left on
short days (Figure 11-1C); under these conditions, maximal gonadal activity
will be maintained indefinitely (Reiter 1969b, 1972). Resumption of sensitivity
to short days requires an intermediate period of exposure to long days of at
least 11 weeks duration (Figure 11-1D; Stetson et al. 1977). In the normal course
of events, exposure to long days of sufficient duration would occur during the
summer months, thus ensuring termination of the refractory state before the
following autumn (Reiter 1972; see Figure 11-1A). Thus, in pragmatic terms,
refractoriness can be defined in two ways: spontaneous recrudescence alone
may be considered to be indicative of the attainment of refractoriness (Figure
11-1B and C), but the refractory condition is also defined by the failure of a
second bout of gonadal regression to occur when the animal is presented with
the normal stimuli for regression (Figure 11-1B, C, and D). Both of these definitions have been exploited in laboratory investigations of the etiology of the
refractory state.

C. Theories for Refractoriness

In general, experiments dealing with the refractory condition in hamsters have
approached the larger question of why hamsters become refractory by at-
tempting to answer subsidiary questions: What is refractoriness? When do
hamsters attain the refractory state? How do hamsters become refractory?
Where is the anatomical site of the refractory condition? There are only two
global theories that attempt to synthesize the answers to these subsidiary questions.

The first theory is a reworking of the gonadostat hypothesis which has been
invoked as an explanation for a number of reproductive events in vertebrates.
According to the gonadostat hypothesis, spontaneous changes in reproductive
activity during short-day treatment can be accounted for by a change in sensitivity to gonadal feedback. In the hamster during the first few weeks of short-
day exposure, increased sensitivity to steroid feedback would participate in
decreasing the secretion of pituitary gonadotropins, thus resulting in gonadal
regression; as short-day exposure continues, a decrease in sensitivity to feedback would allow for gradually increasing secretion of gonadotropins, thus resulting in testicular growth. This gonadostat theory allows for the operation of
signals from neural centers (hypothalamic?) to drive the change in sensitivity.

A very different theory is based on the participation of the pineal gland and
its hormone melatonin in short-day-induced regression. The pineal gland is required for gonadal regression during the initial weeks of short-day treatment,
since pinealectomized hamsters remain fertile on any photoperiod (Czyba et
al. 1964, Hoffman and Reiter 1965, Reiter 1968c). A long line of evidence suggests that the participation of the pineal gland in photoperiod-induced regression
is mediated by the substance melatonin, which is produced within the pineal
in the form of a circadian rhythm, with large amounts produced and secreted
at night and only baseline secretion during the day. Melatonin itself can induce

testicular regression when injected in appropriate quantities in hamsters housed on long days (Tamarkin et al. 1976, Stetson and Tay 1983). The exact mechanism by which melatonin participates in the photoperiodic measurement of daylength is still unknown, but recent evidence indicates that melatonin may act by an internal coincidence mechanism, causing testicular regression when present during the proper phase of a rhythm of sensitivity (Watson-Whitmyre 1985b).

Reiter (1969b, 1972) was the first to suggest that the evidence that the pineal gland (and, later, melatonin itself) was involved in testicular regression could be used to frame a mechanistic theory for the induction of refractoriness. Reiter suggested two alternative ways in which the pineal gland might be involved in the induction of refractoriness: (1) the pineal produces an antigonadotropic substance when the animal is first exposed to short days; exhaustion of the pineal during prolonged short-day treatment allows for recovery of the reproductive system at the time of spontaneous recrudescence; (2) the pineal gland itself continues to produce its antigonadotropin during prolonged short-day treatment, but the target site of the antigonadotropin becomes refractory to its action. An important implication of the latter alternative is that refractoriness to short days is, at a mechanistic level, identical to refractoriness to melatonin.

Virtually all of the experiments dealing with the induction of refractoriness in the hamster have been framed in the context of the gonadostat theory or some version of the pineal theory. It is the purpose of this review to examine the individual experiments and the answers that they have provided to the basic questions of what, when, how, and where refractoriness is produced. Most of the experiments deal with male hamsters because of the relative ease of assaying fertility (by way of gonadal size); the few experiments using female hamsters are included where appropriate. For a review of the role of pineal melatonin in various phases of the annual cycle, see Goldman (1983), Goldman and Darrow (1983), Stetson and Watson-Whitmyre (1984), or Stetson and Watson-Whitmyre (1986).

II. Determinants of Induction of Refractoriness

A. Duration of Short-Day Treatment

Since refractoriness results when hamsters are housed on short photoperiods, one of the first questions asked by investigators was: How many weeks of short-day treatment are necessary for the induction of the refractory state? One of the earliest papers in which Reiter (1972) described the refractory state contains data indicating that hamsters become refractory during the first 10 weeks of short-day treatment. In this experiment, Reiter subjected hamsters to short days (1L:23D) for 10 weeks, then exposed them to long days (14L:10D) for another 10 weeks, to cause photoinduced recrudescence, and finally challenged the hamsters with a second exposure to short days (1L:23D) of 20 weeks duration to test for testicular regression. Gonadal regression ensued during the first 10 weeks, but the hamsters failed to regress during the second exposure to short days, indicating that refractoriness had been attained during the first phase of the experiment. This work by Reiter has been important for two reasons. First, the triphasic experimental paradigm which Reiter introduced has become the standard for research on the induction of refractoriness. Second, an important conclusion from this work is that the refractory state is attained during the initial weeks of short-day treatment, but the outward manifestations of refractoriness (spontaneous recrudescence or failure to regress a second time) are not observed until weeks later.

Zucker and Morin (1977) further investigated the duration of short-day treatment required for the induction of refractoriness. The experimental paradigm was similar to that introduced by Reiter (1972), but Zucker and Morin tested the ability of 3, 4, 6, or 9 weeks of initial short-day exposure (2L:22D) to induce refractoriness. None of the hamsters exposed to 3 or 4 weeks of short days were refractory, as indicated by gonadal regression during a second short-day challenge. Of the hamsters that were exposed to 9 weeks of initial short-day treatment, all but one were refractory to the second short-day challenge. In the group of hamsters exposed to short days for 6 weeks, most failed to demonstrate refractoriness during the second exposure to short days, but two hamsters were refractory, as judged by large testes at the termination of the experiment. Zucker and Morin concluded that 9 weeks of short days appeared to be sufficient to induce refractoriness in a majority of the animals but noted that under normal photoperiodic conditions in the field, the hamsters would be exposed to many months of short days, thus ensuring the attainment of refractoriness simultaneously with complete gonadal regression.

In a series of unpublished studies performed in our laboratory, we examined the minimum duration of short days required for the induction of refractoriness, using the Reiter paradigm of a three-phase experiment, as shown in Figure 11-2a. In the first experiment (Figure 11-2b, top), none of the animals given up to 10 weeks of short days (6L:18D) were refractory to a second short-day exposure of 10 to 12 weeks duration. In the second experiment (Figure 11-2b, middle), 11 to 15 weeks of short days induced refractoriness in some of the animals, but the number of those that achieved refractoriness was still less than 50% in any one group. The third experiment again demonstrated that up to 12 weeks of short days did not cause refractoriness in large numbers of hamsters (Figure 11-2b, bottom). However, Tate-Ostroff and Stetson (1981a) found that as few as 8 weeks of short days induced refractoriness in a significant number of animals; longer durations (10 to 20 weeks) were effective in larger numbers of hamsters, as found also by Zucker and Morin (1977). Our conclusion from all of these studies is that the population of hamsters shows tremendous variability in the duration of short-day exposure necessary to induce refractoriness: it varies from 6 to 20 weeks or more; factors which may affect this duration are currently unidentified.

There have been few studies of the determinants of the induction of refractoriness in female hamsters. Reiter (1968a) originally showed that, like males, females exhibit a spontaneous return to the fertile state during prolonged short-day treatment. Stetson and Anderson (1980) found that resumption of estrous

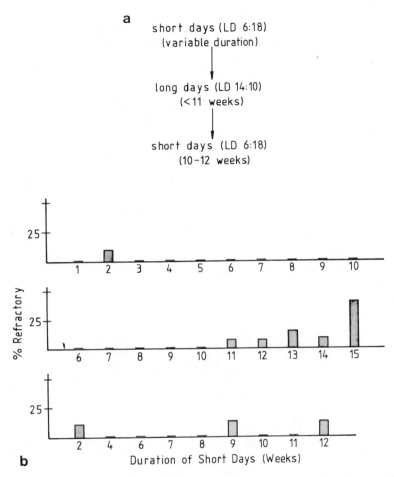

Figure 11-2. Protocol (a) and results (b) of three experiments to determine the minimum duration of short days necessary for the induction of refractoriness. In (b), the shaded bars indicate the percentage of animals judged to be refractory at the end of the experiment, based on a paired testes weight of greater than 2000 mg. Duration of initial short-day treatment is indicated at the bottom of each bar.

cyclicity ensued after an average of 16 weeks of exposure to 6L:18D. Although the exact duration of short-day exposure for the induction of refractoriness in the female hamster is not known, Stetson and Hamilton (1981) found that females that became anovulatory during 4 to 7 weeks of 6L:18D exposure were not refractory to a second exposure to 6L:18D, even though they had remained in the anovulatory state for an average of 6 weeks. Thus, in the female also, it appears that mere cessation of fertility is no guarantee of the induction of refractoriness.

B. Role of the Gonads in the Induction of Refractoriness

1. Role of Testicular Regression

Zucker and Morin (1977) not only examined the duration of short-day treatment necessary for the induction of refractoriness, but also attempted to determine if full testicular regression was a prerequisite to the induction of refractoriness. Three or four weeks of short days did not induce refractoriness in any hamsters and were insufficient for full gonadal regression to occur. Nine weeks of short days induced full regression in all hamsters and refractoriness in most. However, refractoriness and regression were not correlated in the group exposed to 6 weeks of short days: the two animals that were judged to be refractory did not have fully regressed gonads at the end of the initial short-day treatment. Zucker and Morin concluded that complete regression alone is poorly correlated with the induction of refractoriness: some hamsters that do not undergo full regression become refractory, while others that do show full regression fail to become refractory.

In an effort to separate the two inductive effects of short-day treatment (induction of regression versus induction of refractoriness), we designed two experiments to take advantage of the observation that exposure to an exercise wheel often retards gonadal regression (Elliot, Stetson, and Menaker, unpublished; see also Borer et al. 1983). Our prediction was that if the endocrine state of the hamster is important in determining when refractoriness is induced, then hamsters experiencing delayed gonadal regression might require longer durations of short-day exposure to attain refractoriness. As shown in Figure 11-3a, two groups (with running wheels or without) of hamsters were exposed to short days (6L:18D) of from 6 to 14 weeks duration (Experiment 1) or from 6 to 17 weeks (Experiment 2). The results (Figure 11-3b) show that our prediction was not fulfilled: for most durations of short-day treatment, there was no difference between groups in the numbers of hamsters achieving refractoriness; in some groups [e.g., Experiment 2, 9 weeks treatment (Figure 11-3b, bottom)], the hamsters housed on running wheels actually attained refractoriness earlier than those housed in conventional cages. These experiments also show that we were able to induce refractoriness with shorter durations of short-day exposure than had previously been effective (compare Figure 11-2b with Figure 11-3b).

Although the experiments described above lead to the conclusion that full testicular regression per se is neither necessary for the induction of refractoriness nor a good predictor of it, exactly the opposite conclusion was reached in two separate studies employing the hormone melatonin to prevent testicular regression on short days. When given as a subcutaneous implant, melatonin will reliably prevent gonadal regression during short-day treatment (Reiter et al. 1974, 1975, 1976b, Turek et al. 1975a); this has been termed the counterantigonadotropic action of melatonin (Reiter 1982). Bittman (1978b) used this finding to explore the possibility that gonadal regression, rather than short-day exposure per se, is required for the induction of refractoriness. Hamsters were housed

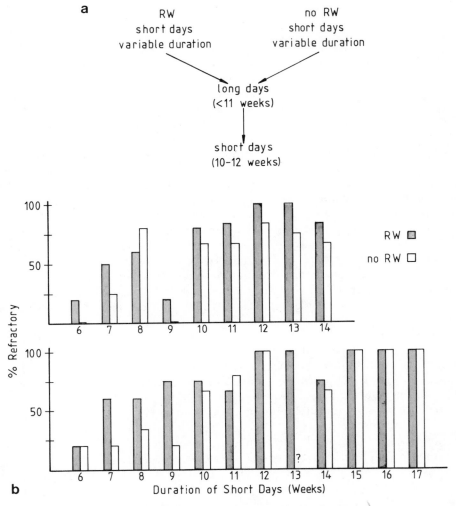

Figure 11-3. Protocol (a) and results (b) of two experiments designed to distinguish between the direct effects of short days and the indirect effects via endocrine changes on induction of refractoriness. In (b), results, the bars show percentage of refractory hamsters (paired testes weight greater than 2000 mg) at the end of the experiment. RW indicates the group had access to running wheels during the initial short-day exposure.

on 2L:22D for 10 or 8.5 weeks; during this period, half of the hamsters bore implants of melatonin. The melatonin-implanted hamsters failed to regress during the initial weeks of short-day treatment, although all of the control hamsters did so. Following the removal of the implants, the animals remained on short days; all exhibited gonadal regression, although the control hamsters, similarly housed on short days, were undergoing spontaneous recrudescence at this time. Bittman concluded that testicular regression itself is necessary for the attainment of refractoriness, since the implanted hamsters regressed when the controls were showing evidence of refractoriness (spontaneous recrudescence).

One question not answered by Bittman's study (1978b) is whether the initial 8.5 or 10 weeks of short-day treatment was truly sufficient to induce refractoriness, if the experimental hamsters had *not* been given melatonin implants. This is a reasonable doubt given the aforementioned variability in the duration of short-day exposure necessary to induce refractoriness. Control hamsters were not included to answer this question. Instead, the assumption was made that since the nonimplanted hamsters were exhibiting spontaneous recrudescence at the same time that the experimentals were first experiencing regression, refractoriness had not been induced in the implanted hamsters. This assumption begs the question of whether the timing of the onset of spontaneous recrudescence is at all related to or influenced by the timing of the initial induction of refractoriness. As will be discussed later, spontaneous recrudescence, as a peripheral phenomenon, appears to be influenced by factors such as gonadal feedback which may not affect the central or initial induction of refractoriness.

A similar study was performed by Turek and Losee (1979), who used a protocol essentially identical to that employed by Bittman, but with longer durations of short-day treatment. In one experiment, implantation with melatonin during the first 24 weeks of short days (6L:18D) prevented the development of refractoriness, as indicated by immediate gonadal regression when the implants were removed; the same response occurred when the implants were removed after only 15 weeks of short days. Like Bittman (1978b), Turek and Losee (1979) concluded that testicular regression is a necessary prerequisite for the induction of refractoriness, as shown in Figure 11-4a. However, the experimental data do not eliminate the possibility, shown in Figure 11-4b, that melatonin implants might prevent testicular regression in short-day-housed hamsters by interfering with the interpretation of daylength, but without a causal connection between regression and induction of refractoriness.

Further support for the idea that testicular regression and the induction of refractoriness are not causally linked comes from a pair of unpublished studies in our laboratory. In the first study, pineal-intact male hamsters were given daily injections of melatonin 1 h after lights off on 14L:10D, a long day (Figure 11-5a). As reported by Stetson and Tay (1983), these injections induce testicular regression followed by recrudescence within 27 weeks (Figure 11-5b). At the end of this time, the hamsters were transferred to constant darkness for 10 weeks and then sacrificed. At the time of sacrifice, only 2 of the 14 hamsters (14%) in the group were refractory, as judged by the failure to show gonadal regression in constant darkness (Figure 11-5b). Thus, in this experiment, regression and subsequent recrudescence were not sufficient to ensure the induction of refractoriness to short days.

In another study, pinealectomized hamsters were given melatonin injections

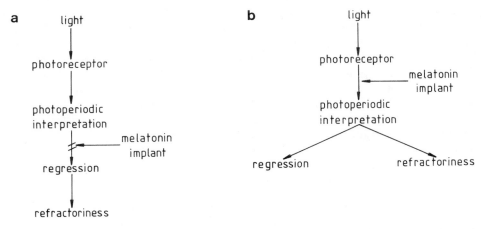

Figure 11-4. Schematic showing two possible mechanisms by which melatonin delivered in a subcutaneous implant may interfere with the induction of refractoriness and regression on short days. In (a), regression is a prerequisite for the induction of refractoriness, and melatonin prevents the induction of refractoriness by interfering with the induction of regression. In (b), refractoriness and regression are separately induced by short days, and melatonin prevents the induction of both by interfering with the mechanism used to discriminate daylength; no causative link exists between the regressed state and the refractory state.

daily 7 hours after lights off in a short-day light cycle (12L:12D; Figure 11-6a) (Watson-Whitmyre and Stetson, 1984a,b). Although the injections failed to induce regression during the 17 weeks of administration, 78% of the hamsters receiving these injections were not responsive to a later 12-week series of three daily melatonin injections (Figure 11-6b), a treatment that reliably induces regression in nonrefractory pinealectomized animals (Tamarkin et al. 1977). Thus, these pinealectomized hamsters had become refractory, at least to melatonin, during the simultaneous treatment with short days and melatonin, even though testicular regression did not occur. There is good evidence (to be discussed later) that refractoriness to short days is indeed equivalent to refractoriness to melatonin; if this is the case, the results of this experiment show that testicular regression (and the concomitant hormonal changes) is not a prerequisite for induction of refractoriness. This conclusion has been upheld in other studies with pinealectomized hamsters that will be described later.

2. Role of Gonadal Feedback

Although the experiments described above demonstrate that the occurrence of testicular regression is not a good predictor for the induction of refractoriness, they do not eliminate the possibility that feedback from the gonads does play a role in the timing of the onset of spontaneous recrudescence, one of the indicators of the successful development of refractoriness. However, it should be noted here that the onset of spontaneous recrudescence is a peripheral event,

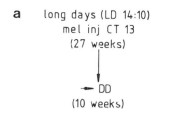

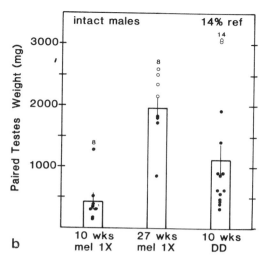

Figure 11-5. Protocol (a) and results (b) of an experiment demonstrating that complete gonadal regression and recrudescence are insufficient for the induction of refractoriness. Bars in (b) show mean paired testes weight estimated from measurements of one testis, with the vertical line indicating the standard error; circles show individual paired testes weights. Numerals at the top of each bar show number of animals per group. Criterion for refractoriness was a final paired testes weight of greater than 2000 mg. Time of injection is shown in circadian time (CT), where CT12 is defined as the time of activity onset; for 14L:10D, CT12 is at the time of lights-off. Other abbreviations used: mel = melatonin ; inj = injected; DD = constant darkness; 1X = one time daily; ref = refractory.

and as such may be influenced by many more factors than is the central induction of refractoriness. Thus, as mentioned earlier, the timing of spontaneous recrudescence is not necessarily strongly linked to the timing of the induction of refractoriness during the initial weeks of short-day exposure. Nevertheless, it has been widely assumed that the two events are temporally related. The experimental design used to investigate this possibility has generally been to expose hamsters to short days for a sufficient duration to exhibit spontaneous recrudescence; groups of hamsters are castrated and/or given testosterone implants at some time during short-day exposure in order to vary the endocrine status of the animals.

The influence of gonadal feedback on the timing of recrudescence was first investigated by Turek et al. (1975b), who measured serum gonadotropin levels in hamsters castrated after 10 weeks exposure to short days (6L:18D). Spontaneous increases in serum gonadotropins were observed in the castrates at the same time that spontaneous recrudescence occurred in the intact controls on the same photoperiod. Turek and colleagues concluded from these results that the increase in serum gonadotropins which accompanies spontaneous recrudescence of the gonads is not dependent upon the testis or feedback therefrom.

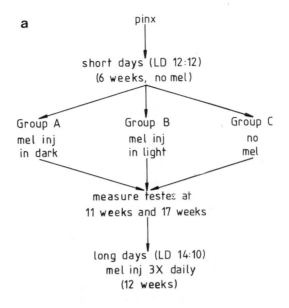

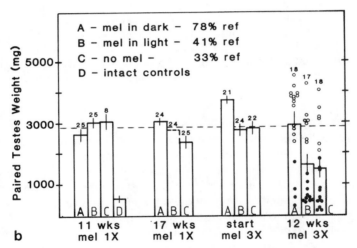

Figure 11-6. Protocol (a) and results (b) of an experiment showing that prior gonadal regression is not a prerequisite for the induction of refractoriness and that the timing of the melatonin signal is important in inducing refractoriness. Bars in (b) show mean paired testes weights with standard error for each group at various time points in the experiment; circles are individual paired testes weights. Numerals at the top of each bar show number of animals per group. Criterion for refractoriness was a paired testes weight of greater than 2000 mg; testes weights were estimated from measurements of one testis. Group D was a group of intact hamsters housed on the same short photoperiod for comparison of the extent of possible regression. Abbreviations used: pinx = pinealectomized; mel = melatonin; inj = injected; 1X = one time daily; 3X = three daily; ref = refractory.

A similar conclusion was reached by Goldman and coworkers (1981), who examined serum prolactin levels in intacts and castrates around the time of spontaneous recrudescence of the gonads during short-day treatment. Tate-Ostroff (1981) examined serum gonadotropin levels in males castrated at the beginning of short-day treatment; her conclusions support those of Turek et al. (1975b) and Goldman et al. (1981) that hormonal changes during spontaneous recrudescence can occur independently of feedback from the gonads.

Other work by Ellis and coworkers (1979) demonstrated that there is also a steroid-dependent component to the gonadotropin changes observed during short-day treatment, since the hypothalamic-pituitary axis of the hamster on short days becomes less sensitive to the negative feedback action of testosterone at about the time of spontaneous recrudescence in intact males. This conclusion was based on experiments in which castrated males were held on short days for prolonged periods and given implants of testosterone; assays of serum gonadotropins indicated that doses of testosterone which were capable of suppressing gonadotropin secretion early in short-day exposure became less able to do so as short-day treatment progressed. Similar results were obtained in a study by Matt and Stetson (1979), who also looked at the ability of testosterone to alter serum gonadotropins in castrated hamsters on short days. Thus, the hypothalamo-pituitary axis undergoes a spontaneous change in sensitivity during the course of short-day treatment: sensitivity to feedback initially increases during the period of gonadal regression, remains high during the quiescent period of the gonads, and then spontaneously decreases at the time of spontaneous recrudescence (Ellis and Turek 1979). Annual changes in feedback sensitivity appear to be mediated by pineal melatonin (Turek 1979, Sisk and Turek 1982, Tate-Ostroff and Stetson 1981b). These changes in feedback sensitivity can help explain the observed cycle of changes in gonadal size and the absence of compensatory changes in serum gonadotropins as short-day exposure progresses. These results further imply that one site for the development of the refractory condition may be the site(s) of feedback action of gonadal steroids at the hypothalamus and pituitary.

Circumstantial support for this hypothesis comes from the work of Tate-Ostroff and Stetson (1981a), who examined the change in feedback sensitivity (as indicated by the magnitude of the postcastration increase in serum gonadotropins) of castrated hamsters on short days and compared this with the timing of the induction of refractoriness. In this study, a minimum of 8 weeks of short days induced refractoriness in a significant number of animals; likewise, the castration response of serum gonadotropins was first attenuated at 8 weeks of short-day treatment. As the authors note, the results from this study are correlational only and do not necessarily show a causative effect of feedback mechanisms in the induction of refractoriness. Nevertheless, the data support the idea that significant alterations in the hypothalamo-pituitary-gonadal axis occur at the same time that refractoriness is being induced by short days.

The involvement of prolactin in the regulation of the annual gonadal cycle has been extensively studied in Bartke's laboratory. In one study (Bartke et al. 1983), it was found that the ability of prolactin (supplied by ectopic pituitary grafts) to influence testicular size paralleled the aforementioned changes in testosterone feedback sensitivity: early in short-day treatment, prolactin causes increases in testicular size (compared to regressing controls); the sensitivity of the testis to prolactin peaks and then decreases during the period of normal spontaneous recrudescence. The ability of prolactin to increase testis size early in short-day treatment is probably due to its effect at two target sites: within the testis, prolactin increases hCG (human chorionic gonadotropin) [luteinizing hormone (LH)] binding (Bartke et al. 1982); and at the pituitary, prolactin may enhance follicle-stimulating hormone (FSH) release (Bartke et al. 1981, 1983). These results point to another possible mechanism for the induction of refractoriness via changes in response to prolactin, at the testis and/or the pituitary.

C. Role of the Pineal in Induction of Refractoriness

1. Is the Presence of the Pineal Gland Necessary for the Maintenance of Refractoriness?

Reiter's early work (1972) on refractoriness indicated that hamsters pinealectomized after either photoinduced or spontaneous recrudescence failed to exhibit gonadal regression when exposed to short days. Since pinealectomy normally prevents gonadal regression on short days, it was impossible to judge whether or not these hamsters were truly refractory to the short photoperiods. However, once Tamarkin and coworkers (1977) demonstrated that nonrefractory pinealectomized hamsters could be made to regress with three daily injections of melatonin, a procedure became available for testing the responsiveness of pinealectomized hamsters; the procedure is based on the assumption that melatonin injections and short days cause regression by the same mechanism, and that refractoriness to melatonin is equivalent to refractoriness to short days (see below). Bittman and Zucker (1981) used triple melatonin injections to test for the ability of pinealectomized hamsters to terminate refractoriness on long days. As shown in Figure 11-7, hamsters were exposed to short days (10L:14D) for a sufficient period to cause gonadal regression and subsequent recrudescence (30 weeks or more). One group was then pinealectomized and immediately given triple injections during 7 weeks of long days; this group failed to regress in response to the injections (Figure 11-7a), showing that refractoriness to short days also resulted in refractoriness to melatonin and that this refractoriness was maintained after the removal of the pineal. Another group was not pinealectomized until after 20 weeks of long-day treatment (with no melatonin injections); this group did respond to a later 7 weeks course of melatonin injections with gonadal regression (Figure 11-7b), thus indicating that responsiveness to melatonin is restored by the same treatment (20 weeks of long days) that restores sensitivity to short days. Pinealectomy at the beginning of the 20 weeks of long days prevented the termination of refractoriness to melatonin (Figure 11-7c); thus the presence of the pineal is required during the period of long-day exposure.

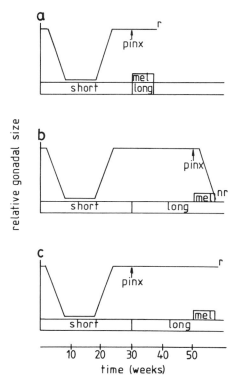

Figure 11-7. Protocol and results for three groups of hamsters in an experiment demonstrating the need for the pineal gland during termination of refractoriness, and that the pineal is not required for the continued expression of refractoriness, once it is induced. The photoperiodic treatment is indicated by the horizontal bars at the bottom of each panel. Time of pinealectomy (pinx) is shown by the vertical arrow. A period of three daily melatonin injections for 7 weeks is shown by the horizontal bars labeled *mel*. The outcome for each group, refractory (r) or nonrefractory (nr), is indicated at the right of each panel. (Adapted from Bittman and Zucker 1981.)

The results of Bittman and Zucker (1981) were confirmed in an experiment of similar design performed in our laboratory (Stetson et al. 1983). In addition, we showed that long-term absence of the pineal gland itself does not prevent gonadal regression in response to the presentation of three daily injections of melatonin, a necessary control for the interpretation of both Bittman and Zucker's (1981) and our own results. The net conclusion from these experiments is that the refractory condition, once induced, persists in the absence of the pineal gland; thus the pineal gland itself is not the site of refractoriness. The experiments also provide good evidence that refractoriness to melatonin is identical to refractoriness to short days, and that the refractory condition for each is terminated by exposure to long days, which requires the presence of the pineal gland. Interestingly, melatonin treatment during long-day exposure may interfere with the successful termination of refractoriness (Losee and Turek 1981, Stetson et al. 1983).

2. The Pineal Gland and the Induction of Refractoriness

An entirely separate question is whether the pineal gland must be present during the induction of refractoriness in the initial weeks of short days. Reiter (1969b, 1972) was the first to formulate hypotheses concerning the possible role of the pineal in the induction of refractoriness. Reiter's hypotheses had two alternative forms. The first stated that the pineal produces an antigonadotropic substance which causes regression during short-day treatment, but that with continued exposure to short days, the pineal antigonadotropin is exhausted, and thus the pineal can no longer suppress the reproductive system and spontaneous recrudescence ensues. The alternative form of this hypothesis is that the pineal gland itself does not appreciably alter its activity during short days, but the neuroendocrine axis becomes sensitive to the pineal antigonadotropin only during short days. Further short-day treatment renders the hypothalamo-pituitary-gonadal axis again insensitive to the pineal antigonadotropin. Reiter (1972) predicted that the pineal exhaustion theory was less likely than the target site sensitivity theory to be tenable; as it has become clear that the pineal antigonadotropin is probably melatonin, further research has supported Reiter's prediction.

The simplest finding in support of an alteration of target site sensitivity is the description by Rollag and associates (1980) that the pineal melatonin rhythm remains remarkably undisturbed throughout all phases of the reproductive cycle. There is no evidence that the pineal's capacity to produce melatonin diminishes during extended exposure to short days.

Even more important is Bittman's study (1978a) demonstrating that pineal-intact hamsters become refractory to melatonin itself. Bittman injected melatonin once daily, shortly before lights out, into hamsters housed on long days; this regimen had previously been shown to induce regression in pineal-intact, but not pinealectomized, hamsters (Reiter et al. 1976a, Tamarkin et al. 1976). Bittman found that the melatonin injections caused testicular regression, as expected; when continued for a longer duration, the injected hamsters underwent testicular recrudescence with a time course similar to that of control hamsters on short days. Furthermore, pineal-intact hamsters whose gonads had regressed and recrudesced during short-day treatment did not regress again when injected with melatonin just before lights out. These experiments provide strong evidence that refractoriness to melatonin is induced by short-day exposure and lend support to the hypothesis that recrudescence of the testes results from the development of refractoriness to melatonin. The data provide strong support for the theory that refractoriness to melatonin and refractoriness to short days are essentially equivalent.

Other support for the idea that refractoriness to melatonin and short days are equivalent comes from the study of Reiter and coworkers (1979), who found that hamsters that had undergone a cycle of regression and recrudescence in response to melatonin injections were refractory to a subsequent exposure to short days. The study by Reiter's group further found that refractoriness to melatonin, like refractoriness to short days, could be terminated by a period of long-day exposure.

Although the work by Bittman (1978a) showed that the target site sensitivity hypothesis originally forwarded by Reiter (1969b, 1972) could account for the development of the refractory condition, it was still not known if the pineal gland itself is necessary for the induction of refractoriness, apart from the role of the pineal in producing melatonin. To answer this question, we injected pinealectomized hamsters with melatonin for a sufficient duration to determine if spontaneous recrudescence would result in the face of continuing melatonin treatment, as it does in pineal-intact hamsters (Bittman 1978a). Hamsters were pinealectomized and housed on long days (14L:10D) (Watson-Whitmyre and Stetson 1985). Melatonin injections were given either three times daily, three hours apart, or twice daily, one at the time of the usual peak of melatonin production within the pineal and one just before lights out. The latter schedule was designed to mimic the melatonin profile in a pineal-intact hamster receiving melatonin injections in the evening; gonadal regression results within 10 to 15 weeks when pinealectomized hamsters are injected with melatonin in this manner (Watson-Whitmyre and Stetson 1983). The effectiveness of triple melatonin injections in causing regression in pinealectomized hamsters was already known from the work of Tamarkin (Tamarkin et al. 1977). Both injection schedules resulted in gonadal recrudescence after 15 to 20 weeks; the timing of regression and recrudescence in the double-injected pinealectomized hamsters was very similar to that observed in intact controls housed on short days (Watson-Whitmyre and Stetson 1985). Recrudescence in the triple-injected hamsters occurred somewhat earlier than normal. Thus, melatonin alone is capable of causing a full cycle of gonadal regression and recrudescence in pinealectomized hamsters; this finding argues that no other pineal factor is necessary for the onset of refractoriness to melatonin.

Other evidence that the pineal gland itself is not required for the development of refractoriness to melatonin is our previously described finding that short-day-housed pinealectomized hamsters given melatonin once daily, during the dark period, become refractory to melatonin, even when testicular regression does not occur (Figure 11-6b, Group A) (Watson-Whitmyre and Stetson 1984a,b). Surprisingly, we have also found that a certain small, but significant, percentage of pinealectomized hamsters housed on short days becomes refractory to melatonin, even in the absence of any melatonin treatment during the short-day exposure. We suspected that short days might directly cause refractoriness to melatonin because in the experiment shown in Figure 11-6, the melatonin injections given during the dark period failed to cause regression, although other experiments underway in our laboratory had led us to expect gonadal regression during this treatment. In this same experiment, about 30% of the pinealectomized hamsters in Group C were refractory to melatonin injections, even though they received no melatonin during the initial short-day exposure (Figure 11-6b). This finding led us to hypothesize that the failure of

the Group A hamsters to regress during the first 17 weeks of the experiment
may have been due to the exposure of all the hamsters to short days for 6 weeks
following pinealectomy before the melatonin injections were begun: Had the
hamsters become refractory to melatonin during this 6-week exposure to short
days?

To test this possibility, we used the protocol shown in Figure 11-8a. Nonrefractory hamsters originally housed on long days were pinealectomized and
then transferred to short days (12L:12D); after 6 weeks of short-day exposure,
they were moved to long days (14L:10D) and given 12 weeks of triple melatonin
injections, which reliably cause regression in nonrefractory hamsters (Tamarkin
et al. 1977). Most of the hamsters in the group responded to the injections with
regression, but a significant number (42%) did not regress (Figure 11-8b), indicating that responsiveness to melatonin had been lost during the initial 6 weeks
of short-day exposure. Further work with this phenomenon has shown that
extension of the short-day exposure to 15 weeks fails to increase the number
of pinealectomized hamsters that become refractory to melatonin in the absence
of any melatonin treatment during short days: in various experiments, the percentage of refractory pinealectomized hamsters has stayed constant at about
30 to 40%. This implies that in a certain percentage of the hamster population,
the refractory condition can be induced by short-day exposure directly; this
effect probably is via a neural route that does not necessarily include the pineal
gland.

3. Mechanism of the Induction of Refractoriness by Pineal Melatonin

Recognition of the role of the pineal and particularly the hormone melatonin
in the induction of refractoriness has led us to ask: What aspect of the melatonin
signal is most important for the development of refractoriness? The results

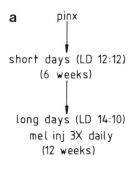
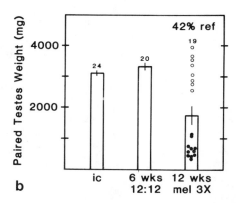

Figure 11-8. Protocol (a) and results (b) for an experiment demonstrating the induction
of refractoriness to melatonin by short days in the absence of the pineal gland. Most
abbreviations, experimental design, and graph layout are the same as for Figure 11-6;
ic = initial controls.

shown in Figure 11-6b demonstrate that some hamsters can become refractory to melatonin in the absence of their own pineal or any exogenous melatonin treatment during short days (Group C). However, the addition of a nightly injection of melatonin during the initial short-day exposure significantly increases the number of hamsters that become refractory to a later challenge with melatonin (Group A). Importantly, melatonin injections given daily during the light period do not increase the number of hamsters that become refractory (Group B). This implies that the timing of the melatonin peak in the circulation may be of paramount importance in the induction of refractoriness (Watson-Whitmyre and Stetson 1984a,b).

To investigate this possibility further, we held pinealectomized golden hamsters on short days (12L:12D) and injected groups at each hour of the light cycle. As reported previously (Watson-Whitmyre and Stetson 1984c, Watson-Whitmyre 1985b), injections given in the first 7 hours of the dark period induce testicular regression within the first 15 weeks of treatment; injections at other times of day have little effect on gonadal weight. The dosage of melatonin used (15 μg) produces circulating levels of melatonin that are greater than those due to endogenous production by the pineal gland, but the melatonin is cleared from the circulation within 2 to 4 hours and clearance does not vary with time of injection (Watson-Whitmyre et al. 1986; Watson-Whitmyre, submitted). Although supraphysiological, this 15-μg dose is still smaller than that originally used by Tamarkin et al. (1976) to demonstrate the existence of a rhythm of sensitivity to melatonin in pineal-intact hamsters. Tamarkin et al. (1976) and Reiter et al. (1976a) had initially reported that single daily injections of melatonin do not cause regression in pinealectomized hamsters; this conclusion was based on the fact that injections in the late evening (before lights out on long days) are ineffective, a finding that has been widely replicated (Tamarkin et al. 1977, Bittman 1978a, Watson-Whitmyre and Stetson 1983). However, all of these studies failed to inject the pinealectomized hamsters during the period of sensitivity identified in the present study (from lights out to 7 hours later, on *short days*).

We hypothesized that refractoriness, like regression, might depend on the interaction of melatonin with a sensitive target site at the appropriate time, and that sensitivity to melatonin might be activated by short-day exposure. To determine if the induction of refractoriness is dependent on a rhythm of sensitivity to melatonin, pinealectomized hamsters that had been housed on short days for 15 weeks while receiving daily melatonin injections at one of 24 times during the light–dark cycle were treated as shown in Figure 11-9a. Hamsters that underwent gonadal regression during the initial 15 weeks of melatonin injection were housed on short days for an additional 15 weeks of injection, to cause recrudescence; they were then transferred to long days (14L:10D) and given three daily injections of melatonin for 15 weeks, to test for refractoriness induced in the first phase of the experiment. The animals whose testes did not regress in response to the first 15 weeks of melatonin injection were immediately transferred to long days (14L:10D) and given three daily injections of melatonin. The results, depicted as percentage of refractory animals in each group (Figure

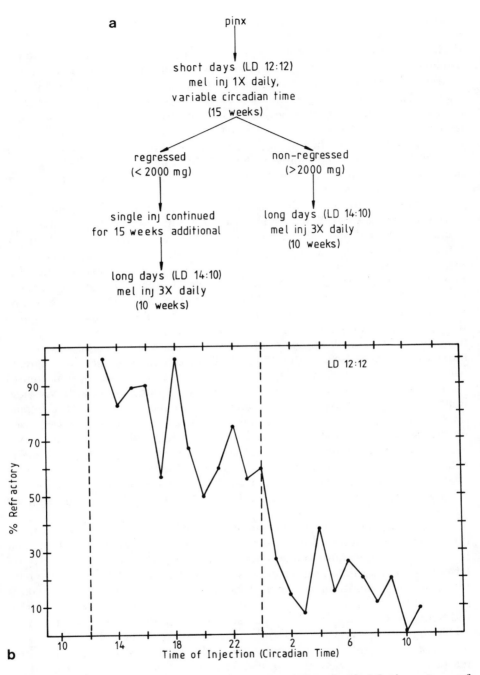

Figure 11-9. Protocol (a) and results (b) for an experiment showing the importance of the time of day that the melatonin signal is presented for the induction of refractoriness on short days. The results of the two experimental treatments shown in the protocol are combined in b according to the time of the single melatonin injection during initial short-day treatment, regardless of whether that injection caused regression. Abbreviations are those used in previous figures.

11-9b), show that hamsters that received the single daily melatonin injections during the dark period were much more likely to develop refractoriness than animals injected in the light. Statistical analysis of the individual results showed that regression itself was a poor predictor of development of refractoriness, but time of injection was very strongly correlated with induction of refractoriness.

In a further test of the mechanism by which melatonin induces refractoriness, we repeated the experiment shown in Figure 11-9, but the hamsters were housed on long days (14L:10D) for the initial 15 weeks of single daily melatonin injections. Only hamsters injected 2 hours after lights out experienced significant gonadal regression, and this regression was much delayed, compared to intact controls or melatonin-treated pinealectomized hamsters on short days (Watson-Whitmyre 1985a,b). After the initial 15 weeks of injection, the hamsters were given a course of three daily melatonin injections, to test for development of refractoriness. As shown in Figure 11-10, there was no apparent rhythm in the response of hamsters to single injections of melatonin during long-day exposure: injections during the dark were no more likely than those during the light to cause induction of refractoriness. Combined with the results shown in Figure 11-9, these results strongly suggest that it is not melatonin exposure per se which causes the development of the refractory condition, but rather the interaction of melatonin with a target site which varies its sensitivity in a rhythmic fashion. Furthermore, the sensitivity rhythm is altered by photoperiod, being activated on short days and damped on long days. Since gonadal regression is not always correlated with the development of refractoriness, it is probable that melatonin acts at more than one target site (i.e., to cause regression or

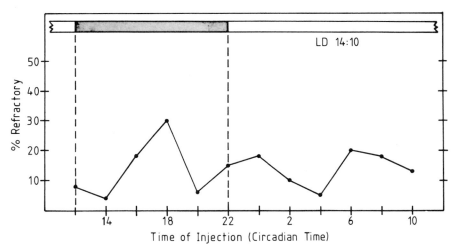

Figure 11-10. Results of an experiment similar to that of Figure 11-9, with the exception that the hamsters were housed on long days (14L:10D) during the initial period of single melatonin injections. See text for further explanation.

refractoriness). The results suggest that perhaps refractoriness results from a disturbance in the normal rhythm of sensitivity to melatonin for the induction of gonadal regression.

4. Target Site for Melatonin

Most of the work that has been done on possible target sites for melatonin have studied the role of melatonin in the induction of testicular regression rather than refractoriness. In general, most workers agree that melatonin's most likely site of action is neuronal, probably within the hypothalamus (Glass and Lynch 1982a,b, Reiter et al. 1981). Our results described above are circumstantial evidence for a neuronal site of action for melatonin in the induction of refractoriness, since photoperiod appears to directly alter the sensitivity of the animals to the inductive action of melatonin: pinealectomized hamsters given melatonin during the dark period of short days are more likely to develop refractoriness than pinealectomized hamsters given melatonin during the dark period of long days. The implication is that the photoperiod changes the response of the animal to melatonin, in the absence of the pineal gland. The most probable route by which photoperiodic information could affect target site sensitivity in the absence of the pineal is a neural route. Definitive studies on the neural basis for the induction of refractoriness have yet to be reported. In this regard, Steger et al. (1985) have shown that at least one neuroendocrine change associated with short-day treatment (a reduction in norepinephrine turnover in the medial basal hypothalamus) is not altered by previous pinealectomy.

5. Pinealectomy-Induced Recrudescence

Still another means of testing for the involvement of the pineal and its product melatonin in the development of refractoriness was suggested by the observation of Reiter (1969b; see also Turek and Ellis 1980) that pinealectomy of hamsters housed on short days after gonadal regression is complete results in almost immediate gonadal recrudescence, far in advance of the usual spontaneous recrudescence seen on short days. On the other hand, pinealectomy at the beginning of long-day exposure does not interfere with photoperiod-induced recrudescence in hamsters with regressed gonads due to prior short-day treatment (Turek 1977, Matt and Stetson 1980, Rusak 1980). Recrudescence of the gonads in pinealectomized hamsters (on short or long days) is similar in rate and hormonal correlates to that seen in pineal-intact hamsters undergoing photoinduced gonadal recrudescence (Matt and Stetson 1980). Several investigators have examined recrudescence in intact or pinealectomized hamsters on both long and short days to determine whether melatonin can modulate the timing of the regrowth of the gonads under these conditions.

Reiter et al. (1976c) first observed that the introduction of melatonin implants to hamsters with regressed testes did not interfere with testicular regrowth: the implanted hamsters attained testes sizes comparable to those of photostimulated hamsters, even though the implanted animals were housed on short days. Reiter et al. (1978) and Turek and Losee (1978) further found that melatonin implants actually stimulated gonadal growth in pineal-intact hamsters housed on short days, compared to control hamsters housed on the same photoperiod. However, if the implanted hamsters were transferred to long days at the time of implantation, gonadal recrudescence was delayed, compared to same-photoperiod controls (Turek et al. 1976, Turek 1977, Turek and Losee 1978). Turek and Losee concluded that the effect of melatonin on gonadal growth is determined by the ambient photoperiod. These authors further noted that, on any photoperiod, melatonin cannot prevent gonadal growth indefinitely, but can only delay it.

Turek (1977) also studied the dependence of the melatonin-induced delay in photostimulated gonadal growth on the presence of the pineal. Hamsters were placed on short days (6L:18D) to induce gonadal regression and were then pinealectomized and implanted with melatonin. Controls were also implanted with melatonin. Both the controls and the pinealectomized hamsters were next transferred to long days (14L:10D) to induce gonadal recrudescence. The effectiveness of melatonin implants in delaying regrowth of the testes was not obstructed by pinealectomy. This finding was verified by Rusak (1980), using a similar experimental design. Rusak also studied photoinduced recrudescence in hamsters bearing lesions of the suprachiasmatic nucleus (SCN) of the hypothalamus. The SCN is reported to be a circadian oscillator, affecting a multitude of circadian rhythms in rodents (Stetson and Watson-Whitmyre 1976, Rusak and Morin 1976, Rusak 1977, Watson-Whitmyre and Stetson 1977); one of the rhythms driven by the SCN is the pineal rhythm of melatonin production (Moore and Klein 1974). Rusak (1980) hypothesized that SCN-lesioned and pinealectomized hamsters should be equivalent in their responses to photoperiod and melatonin treatment if the only (photoperiodically relevant) effect of lesioning the SCN is to disrupt the pineal melatonin rhythm, as some workers had theorized. However, Rusak found that SCN-lesioned and pinealectomized hamsters were not the same in their responses to melatonin: melatonin implants could not delay gonadal growth in SCN-lesioned hamsters moved to long days. Furthermore, SCN lesions caused more rapid gonadal growth than in control hamsters, even in the absence of melatonin implants. Rusak concluded that SCN lesions may disrupt neural circuits that are important in modulating gonadal activity, or that the SCN may be a target site for the action of melatonin. The latter hypothesis is not consistent with the findings of Bittman et al. (1979), who concluded that the SCN is not an essential target for the action of melatonin in inducing regression. However, it should be noted that Bittman and coworkers and Rusak studied the effects of melatonin during two different phases of the hamster's reproductive cycle: data from our laboratory (discussed above) have suggested that melatonin may indeed have different sites of action in the induction of regression and refractoriness.

One question not answered by the published literature is whether the hastened recrudescence seen in pineal-intact hamsters housed on short days with melatonin implants is also seen in pinealectomized hamsters given melatonin. To answer this question, we pinealectomized hamsters after 10 weeks of short-day (6L:18D) exposure, when the gonads were regressed, and then gave the animals

either empty or melatonin-filled capsules, or started a series of three daily melatonin injections (Figure 11-11a) (Watson-Whitmyre and Stetson, unpublished).
Half of the hamsters remained on short days; the other half were transferred
to long days (14L:10D). Figure 11-11b shows that, as expected, pinealectomized
hamsters remaining on short days exhibited gonadal growth sooner than controls. Figure 11-11c shows that melatonin implants did not significantly delay
testes growth in pinealectomized hamsters on short days, but that melatonin
injections were effective in doing so. The results on long days (Figure 11-11d)

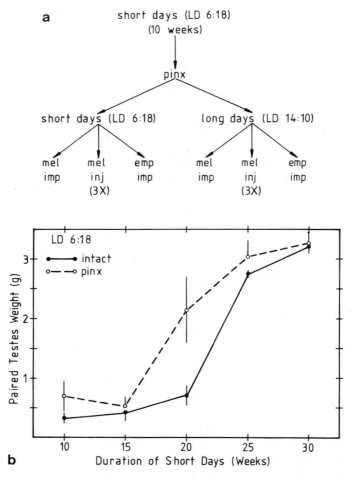

Figure 11-11. Protocol (a) and results (b, c, and d) of an experiment showing the effects
of melatonin implants or injections on premature recrudescence induced by pinealectomy.
The basic effect of pinealectomy of hamsters with regressed gonads is shown in (b).
The effect of introducing melatonin implants or three daily melatonin injections to pinealectomized hamsters housed on short days is shown in (c); the effect on long days
is shown in (d). Abbreviations not used previously: emp = empty; imp = implant.

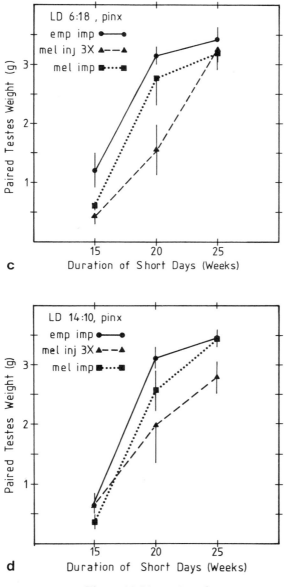

Figure 11-11 *continued.*

verified Turek's (1977) finding that melatonin implants delay photoinduced growth in both pinealectomized and intact hamsters; this effect is even more noticeable with melatonin injections.

These results show that melatonin implants which promote gonadal growth in intact hamsters on short days do not do so in pinealectomized hamsters. Thus the pinealectomized hamster undergoing gonadal recrudescence is still sensitive to the antigonadal effects of melatonin, whether the ambient photo-period is long or short. The implication is that although pinealectomy at 10 weeks of short-day exposure induces almost immediate gonadal growth, these hamsters are not refractory, at least to melatonin. In intact hamsters on short days, melatonin implants might cause premature gonadal growth by acting as a functional pinealectomy, perhaps by interfering with the action of pineal melatonin at its target site.

III. Summary

The experiments described above have provided some important insights into the basic questions about the refractory condition in the golden hamster. Refractoriness to short days appears to be equivalent to refractoriness to melatonin. The refractory condition is apparently induced during the initial weeks of short-day treatment, concomitantly with gonadal regression, but not necessarily causally linked to regression. Gonadal feedback appears to be unimportant for the development of refractoriness, but may influence the timing of spontaneous or photoinduced recrudescence.

The refractory condition may be induced as a result of an interaction between melatonin of pineal origin and a rhythm of target site sensitivity, which is activated by short days. It is plausible that, at a mechanistic level, refractoriness results from the interaction of a neural "refractoriness center" and a "regression center" (Figure 11-12); both of these centers are postulated to be target sites for melatonin, where receptor availability varies in a rhythmic fashion. Short days, acting through the primary oscillator in the SCN, cause the rhythm in melatonin sensitivity within the "regression center" to be activated, thus resulting in gonadal regression as gonadotropin levels begin to decline. At the same time, the rhythm of melatonin sensitivity within the "refractoriness center" is turned on, and refractoriness to short days (melatonin) is induced as melatonin interacts with rhythmically available receptors. The lag period between the initial induction of refractoriness and the onset of spontaneous recrudescence may be due to peripheral factors and to gonadal feedback. Eventually, the insensitivity that is activated in the "refractoriness center" acts on the "regression center" to reverse gonadal regression and result in recrudescence. The "refractoriness center" would continue to dominate the "regression center" until sensitivity to melatonin is restored by long-day treatment.

Support for such a model must come from extensive neuroanatomical and neuroendocrinological investigations. Much work remains to be done to determine the neuroendocrine correlates of the development of refractoriness and the initiation of gonadal recrudescence. However, it is imperative that future work distinguish between recrudescence, which is merely a symptom of refractoriness, and the actual neural switch to the refractory state.

Acknowledgments. We would like to thank Kathy Olsen and Barbara Tate-Ostroff for technical assistance. Experiments performed in our laboratory were supported by a University of Delaware Research Foundation grant and NSF Grants PCM75-04039, PCM78-06664, PCM81-11384, and DCB84-12587 to M.H.S.

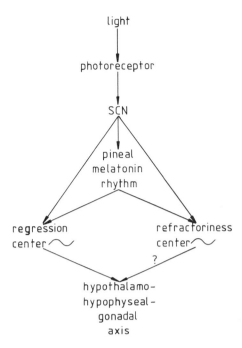

Figure 11-12. A model for the mechanism of the induction of refractoriness by short days in hamsters. The model is constructed to take into account all data to date. Light in the form of a daily photoperiod acts on a photoreceptor, which is known to be the eye. The eye transmits photoperiodic information to the SCN, which acts as a neural oscillator for many rhythms and is known to be a component of the photoperiodic response mechanism. One of the rhythms driven by the SCN is the rhythm of melatonin production in the pineal. The model proposes that two rhythms (indicated by sine waves) of sensitivity to melatonin are also driven by the SCN, one in a proposed "regression center" and one in a proposed "refractoriness center." Short days activate the rhythm of sensitivity to melatonin in the "regression center," causing gonadal regression through actions on the hypothalamo-hypophyseal-gonadal axis. Short days also activate the rhythm of melatonin sensitivity in the "refractoriness center." The "refractoriness center" is eventually able to overcome inhibitory signals emanating from the "regression center," thereby resulting in gonadal recrudescence and subsequent refractoriness to further short-day treatment. Long days would terminate refractoriness by damping the rhythm of melatonin sensitivity in the "refractoriness center." The current data do not demand that the "refractoriness center" has direct input to the hypothalamo-hypophyseal-gonadal axis; thus this arrow is marked with a question mark.

References

Bartke A, Siler-Khodr TM, Hogan MP, Roychoudhury P (1981) Ectopic pituitary transplants stimulate synthesis and release of follicle-stimulating hormone in golden hamsters. Endocrinology 108:133–139.

Bartke A, Klemcke HG, Amador A, VanSickle M (1982) Photoperiod and regulation of gonadotropin receptors. Ann NY Acad Sci 383:122–134.

Bartke A, Klemcke HG, Amador A, Goldman BD, Siler-Khodr TM (1983) Relationship of the length of exposure to short photoperiod to the effect of prolactin on pituitary and testicular function in the golden hamster. J Reprod Fert 69:587–595.

Berndtson WE, Desjardins C (1974) Circulating LH and FSH levels and testicular function in hamsters during light deprivation and subsequent photoperiodic stimulation. Endocrinology 95:195–205.

Bittman EL (1978a) Hamster refractoriness: the role of insensitivity of pineal target tissues. Science 202:648–650.

Bittman EL (1978b) Melatonin prevents refractoriness to short days in male hamsters. Proc Soc Exp Biol Med 158:359–362.

Bittman EL, Zucker I (1981) Photoperiodic termination of hamster refractoriness: participation of the pineal gland. Biol Reprod 24:568–572.

Bittman EL, Goldman BD, Zucker I (1979) Testicular responses to melatonin are altered by lesions of the suprachiasmatic nuclei in golden hamsters. Biol Reprod 21:647–656.

Borer KT, Campbell CS, Tabor J, Jorgenson K, Kandarian S, Gordon L (1983) Exercise reverses photoperiodic anestrus in golden hamsters. Biol Reprod 29:38–47.

Czyba J, Girod C, Durand N (1964) Sur l'antagonisme epiphysio-hypophysaire et les variations saisonnières de la spermatogénèse chez le hamster doré (Mesocricetus auratus). C R Soc Biol 158:742–745.

Ellis GB, Turek FW (1979) Time course of the photoperiod-induced change in sensitivity of the hypothalamic-pituitary axis to testosterone feedback in castrated male hamsters. Endocrinology 104:625–630.

Ellis GB, Losee SH, Turek FW (1979) Prolonged exposure of castrated male hamsters to a nonstimulatory photoperiod: spontaneous change in sensitivity of the hypothalamic-pituitary axis to testosterone feedback. Endocrinology 104:631–635.

Gaston S, Menaker M (1967) Photoperiodic control of hamster testis. Science 158:925–928.

Glass JD, Lynch GR (1982a) Evidence for a brain site of melatonin action in the white-footed mouse, Peromyscus leucopus. Neuroendocrinology 34:1–6.

Glass JD, Lynch GR (1982b) Diurnal rhythm of response to chronic intrahypothalamic melatonin injections in the white-footed mouse, Peromyscus leucopus. Neuroendocrinology 35:117–122.

Goldman BD (1983) The physiology of melatonin in mammals. Pineal Res Rev 1:145–182.

Goldman BD, Darrow JM (1983) The pineal gland and mammalian photoperiodism. Neuroendocrinology 37:386–396.

Goldman BD, Matt KS, Roychoudhury P, Stetson MH (1981) Prolactin release in golden hamsters: photoperiod and gonadal influences. Biol Reprod 24:287–292.

Hoffman RA, Reiter RJ (1965) Pineal gland: influence on gonads of male hamsters. Science 148:1609–1611.

Losee SH, Turek FW (1981) Melatonin treatment prevents the termination of the gonadal-refractory conditions normally observed in hamsters exposed to long days. In: Matthews CD, Seamark RF (eds) Pineal Function. Elsevier/North-Holland, Amsterdam, pp. 67–76.

Matt KS, Stetson MH (1979) Hypothalamic-pituitary-gonadal interactions during spontaneous testicular recrudescence in golden hamsters. Biol Reprod 20:739–746.

Matt KS, Stetson MH (1980) A comparison of serum hormone titers in golden hamsters during testicular growth induced by pinealectomy and photoperiodic stimulation. Biol Reprod 23:893–898.

Moore RY, Klein DC (1974) Visual pathways and the central neural circadian rhythm in pineal serotonin *N*-acetyltransferase activity. Brain Res 71:17–33.

Reiter RJ (1968a) Changes in the reproductive organs of cold-exposed and light-deprived female hamsters *(Mesocricetus auratus)*. J Reprod Fertil 16:217–224.

Reiter RJ (1968b) Morphological studies on the reproductive organs of blinded male hamsters and the effects of pinealectomy or superior cervical ganglionectomy. Anat Rec 160:13–24.

Reiter RJ (1968c) The pineal gland and gonadal development in male rats and hamsters. Fertil Steril 19:1009–1017.

Reiter RJ (1969a) Pineal function in long term blinded male and female golden hamsters. Gen Comp Endocrinol 12:460–468.

Reiter RJ (1969b) Pineal-gonadal relationships in male rodents. In: Gual C (ed) Progress in Endocrinology. Excerpta Medica, Amsterdam, pp. 631–636.

Reiter RJ (1972) Evidence for refractoriness of the pituitary-gonadal axis to the pineal gland and its possible implications in annual reproductive rhythms. Anat Rec 173:365–372.

Reiter RJ (1980) The pineal and its hormones in the control of reproduction in mammals. Endocr Rev 1:109–131.

Reiter RJ (1982) Neuroendocrine effects of the pineal gland and of melatonin. In: Ganong WF, Martini L (eds) Frontiers in Neuroendocrinology, Vol. 7. Raven Press, New York, pp. 287–316.

Reiter RJ, Vaughan MK, Blask DE, Johnson LY (1974) Melatonin: its inhibition of pineal antigonadotropic activity in male hamsters. Science 185:1169–1171.

Reiter RJ, Vaughan MK, Waring PJ (1975) Studies on the minimal dosage of melatonin required to inhibit pineal antigonadotropic activity in male golden hamsters. Horm Res 6:258–267.

Reiter RJ, Blask DE, Johnson LY, Rudeen PK, Vaughan MK, Waring PJ (1976a) Melatonin inhibition of reproduction in the male hamster: its dependency on time of day of administration and on intact and sympathetically innervated pineal gland. Neuroendocrinology 22:107–116.

Reiter RJ, Rudeen PK, Vaughan MK (1976b) Restoration of fertility in light-deprived female hamsters by chronic melatonin treatment. J Comp Physiol 111:7–13.

Reiter RJ, Vaughan MK, Rudeen PK, Philo RC (1976c) Melatonin induction of testicular recrudescence in golden hamsters and its subsequent inhibitory action on the antigonadotropic influence of darkness on the pituitary-gonadal axis. Am J Anat 147:235–242.

Reiter RJ, Rudeen PK, Philo RC (1978) Influence of chronic melatonin availability on the reproductive quiescent period in male hamsters exposed to natural photoperiods during the winter months. In: Veale WL, Lederis K (eds) Current Studies on Hypothalamic Function. Karger, Basel, pp. 175–182.

Reiter RJ, Petterborg LJ, Philo RC (1979) Refractoriness to the antigonadotropic effects of melatonin in male hamsters and its interruption by exposure of the animals to long daily photoperiods. Life Sci 25:1571–1576.

Reiter RJ, Dinh DT, de Los Santos R, Guerra JC (1981) Hypothalamic cuts suggest a brain site for the antigonadotropic action of melatonin in the Syrian hamster. Neurosci Lett 23:315–318.

Rollag MD, Panke ES, Reiter RJ (1980) Pineal melatonin content in male hamsters throughout the seasonal cycle. Proc Soc Exp Biol Med 165:330–334.

Rusak B (1977) The role of the suprachiasmatic nuclei in the generation of circadian rhythms in the golden hamster. J Comp Physiol 118:145–164.

Rusak B (1980) Suprachiasmatic nucleus lesions prevent an antigonadal effect of melatonin. Biol Reprod 22:148–154.

Rusak B, Morin LP (1976) Testicular responses to photoperiod are blocked by lesions of the suprachiasmatic nuclei in golden hamsters. Biol Reprod 15:366–374.

Sisk CL, Turek FW (1982) Daily melatonin injections mimic the short day-induced increase in negative feedback effects of testosterone on gonadotropin secretion in hamsters. Biol Reprod 27:602–608.

Steger RW, Matt KS, Bartke A (1985) Neuroendocrine regulation of seasonal reproductive activity in the male golden hamster. Neurosci Biobehav Rev 9:191–201.

Stetson MH, Anderson PJ (1980) Circadian pacemaker times gonadotropin release in freerunning female hamsters. Am J Physiol 238:R23–27.

Stetson MH, Hamilton B (1981) The anovulatory hamster: a comparison of the effects of short photoperiod and daily melatonin injections on the induction and termination of ovarian acyclicity. J Exp Zool 215:173–178.

Stetson MH, Tate-Ostroff B (1981) Hormonal regulation of the annual reproductive cycle of golden hamsters. Gen Comp Endocrinol 45:329–344.

Stetson MH, Tay DE (1983) Time course of sensitivity of golden hamsters to melatonin injections throughout the day. Biol Reprod 29:432–438.

Stetson MH, Watson-Whitmyre M (1976) The nucleus suprachiasmaticus: the biological clock in the hamster? Science 191:197–199.

Stetson MH, Watson-Whitmyre M (1984) Physiology of the pineal and its hormone melatonin in annual reproduction in rodents. In: Reiter RJ (ed) The Pineal Gland. Raven Press, Oxford, pp. 109–153.

Stetson MH, Watson-Whitmyre M (1986) Effects of exogenous and endogenous melatonin on gonadal function in hamsters. J Neural Trans (Suppl.) 21:55–80.

Stetson MH, Watson-Whitmyre M, Matt KS (1977) Termination of photorefractoriness in golden hamsters—photoperiodic requirements. J Exp Zool 202:81–88.

Stetson MH, Watson-Whitmyre M, Tate-Ostroff B (1983) The role of the pineal and its hormone melatonin in the termination of photorefractoriness in golden hamsters. Biol Reprod 29:689–696.

Tamarkin L, Westrom WK, Hamill AI, Goldman BD (1976) Effect of melatonin on the reproductive systems of male and female hamsters: a diurnal rhythm in sensitivity to melatonin. Endocrinology 99:1534–1541.

Tamarkin L, Hollister CW, Lefebvre NG, Goldman BD (1977) Melatonin induction of gonadal quiescence in pinealectomized Syrian hamsters. Science 198:953–955.

Tate-Ostroff B (1981) The photoperiodic testicular response of the golden hamster: mechanisms regulating gonadotropin release. Ph.D. dissertation, University of Delaware, Newark.

Tate-Ostroff B, Stetson MH (1981a) Correlative changes in the response to castration and the onset of refractoriness in male golden hamsters. Neuroendocrinology 32:325–329.

Tate-Ostroff B, Stetson MH (1981b) Melatonin-induced suppression of gonadotropin titers in male golden hamsters: effect on gonadal feedback mechanisms. Life Sci 28:243–250.

Turek FW (1977) Antigonadal effect of melatonin in pinealectomized and intact male hamsters. Proc Soc Exp Biol Med 155:31–34.

Turek FW (1979) Role of the pineal gland in photoperiod-induced changes in hypothalamic-pituitary sensitivity to testosterone feedback in castrated male hamsters. Endocrinology 104:636–640.

Turek FW, Ellis GB (1980) Role of the pineal gland in seasonal changes in neuroendocrine-testicular function. In: Steinberger E, Steinberger A (eds) Testicular Development, Structure, and Function. Raven Press, New York, pp. 389–393.

Turek FW, Losee SH (1978) Melatonin-induced testicular growth in golden hamsters maintained on short days. Biol Reprod 18:299–305.

Turek FW, Losee SH (1979) Photoperiodic inhibition of the reproductive system: a prerequisite for the induction of the refractory period in hamsters. Biol Reprod 20:611–616.

Turek FW, Desjardins C, Menaker M (1975a) Melatonin: antigonadal and progonadal effects in male golden hamsters. Science 190:280–282.

Turek FW, Elliott JA, Alvis JD, Menaker M (1975b) Effect of prolonged exposure to non-stimulatory photoperiods on the activity of the neuroendocrine-testicular axis of golden hamsters. Biol Reprod 13:475–481.

Turek FW, Desjardins C, Menaker M (1976) Melatonin-induced inhibition of testicular function in adult golden hamsters. Proc Soc Exp Biol Med 151:502–506.

Vendrely E, Guerillot C, Basseville C, DaLage C (1971a) Poids testiculaire et spermatogénèse du hamster doré au cours du cycle saisonnier. C R Soc Biol 165:1562–1565.

Vendrely E, Guerillot C, Basseville C, DaLage C (1971b) Variations saisonnières de la spermatogénèse chez le hamster doré. Bull Assoc Anat 152:778–788.

Vendrely E, Guerillot C, DaLage C (1972) Variations saisonnières de l'activité des cellules de Sertoli et de Leydig dans le testicule du hamster doré. Etude caryométrique. C R Acad Sci Paris 275:1143–1146.

Watson-Whitmyre M (1985a) Photoperiodic regulation of sensitivity to melatonin in pinealectomized golden hamsters. Biol Reprod 32 (Suppl. 1):175.

Watson-Whitmyre M (1985b) Photoperiodism in the golden hamster: dependence on rhythmic sensitivity to melatonin. Ph.D. dissertation, University of Delaware, Newark.

Watson-Whitmyre M, Stetson MH (1977) Circadian organization in the regulation of reproduction: identification of a circadian pacemaker in the hypothalamus of the hamster. J Interdiscipl Cycle Res 8:360–367.

Watson-Whitmyre M, Stetson MH (1983) Simulation of peak pineal melatonin release restores sensitivity to evening melatonin injections in pinealectomized hamsters. Endocrinology 112:763–765.

Watson-Whitmyre M, Stetson MH (1984a) A single daily injection of melatonin induces refractoriness in pinealectomized hamsters exposed to short daylengths. J Steroid Biochem 20:1474.

Watson-Whitmyre M, Stetson MH (1984b) Induction of refractoriness in male hamsters in the absence of gonadal regression. Biol Reprod 30 (Suppl. 1):165.

Watson-Whitmyre M, Stetson MH (1984c) Single injections of melatonin induce testicular regression in pinealectomized hamsters on short days. The Physiologist 27:227.

Watson-Whitmyre M, Stetson MH (1985) Induction of refractoriness to melatonin in pinealectomized hamsters. In: Brown GM, Wainwright SD (eds) The Pineal Gland: Endocrine Aspects. Pergamon, Oxford, pp. 121–126.

Watson-Whitmyre M, Rollag MD, Stetson MH (1986) Disappearance of melatonin from the serum of pinealectomized golden hamsters following chronic subcutaneous injections. Biol Reprod 34 (Suppl. 1):181.

Watson-Whitmyre M, Rollag MD, Stetson MH. Disappearance of exogenous melatonin from the serum of pinealectomized golden hamsters (submitted).

Zucker I, Morin LP (1977) Photoperiodic influences on testicular regression, recrudescence and the induction of scotorefractoriness in male golden hamsters. Biol Reprod 17:493–498.

Index

A

Acanthobrama terrae-sanctae
  seasonal thyroid activity, 3
  peak thyroid activity, 5
Acipenser stallatus (stellate sturgeon)
  thyroid and gonadal function, 8
Acomys cahirinus (spiny mouse)
  indirect effect of light on fetus, 169
  perinatal communication of circadian
    phase, 154, 168
  precocial development, 168
ACTH (adrencocorticotropin)
  and suppression of reproductive be-
    havior in amphibia, 40
  effects on reproductive organs in
    birds, 134
activity rhythm
  circadian nature, 49
  entrainment, 56, 65
  freerun, 50
  phase shift, 54, 55
adrenocorticotropic hormone, *see* ACTH
adrenocorticotropin, *see* ACTH
α-adrenergic pathways
  effects on LH in sheep, 193
amago salmon *(Onchorhynchus rhodurus)*
  effect of T_4 on smoltification, 4
Ambystoma tigrinum
  seasonal testicular cycle, 22
  serum androgens, seasonal, 25
Amphibians
  ovarian steroids, 26

reproductive behavior 22, 26, 32, 34,
  36, 37, 40
seasonal reproduction, 21–23
  synchronization of, 20
testicular steroids, 24
Anabas testudineus (freshwater perch)
  thyroid and reproductive cycles, 8
Anas platyrhynchos (mallard)
  lack of effect of ACTH and corticos-
    terone on LH, 135
  LH and replacement clutches, 123
androgens
  seasonal rhythms in amphibians, 24–25
androstenedione
  in amphibians, 24, 26
Anguilla vulgaris (eel)
  peak thyroid activity, 5
  seasonal thyroid activity, 2
Anolis carolinensis, 49, 51
  pineal melatonin rhythm, 59, 60, 61, 66
Argenine vasotocin, *see* AVT
arrhythmicity
  in circadian systems, 49
Ascaphus sp., internal fertilization, 20
Aschoff's rule
  definition, 66
  illustration, 64
Astyanax mexicanus (cavefish)
  photoperiod and thyroid function, 8
Atlantic salmon *(Salmo salar)*, 2, 6
AVT
  function in amphibian reproduction,
    33, 37

DATE DUE

AUG 26 1988		
NOV 8 1990		
AUG 31 1991		
MAR 1 6 1992		
MAY 2 6 1992		
JUL 6 1992		
JAN 2 6 1994		
MAY 2 5 1995		
MAY 1 9 1996		

DEMCO NO. 38-298